D0240461

Introduction to Digital Board Testing

R.G. Bennetts

Crane Russak *New York*
Edward Arnold *London*

Introduction to Digital Board Testing
Published in the United States by
   Crane, Russak & Company, Inc.
   3 East 44th Street
   New York, New York 10017
   ISBN 0-8449-1385-0
   LC 81-3258

Published in Great Britain by
   Edward Arnold (Publishers) Ltd.
   41 Bedford Square
   London WC1B 3DQ
   ISBN 0 7131 3450 X

Printed in the United States of America

Contents

Contents

Contents

Mark, Kevin, and Helen

Preface

Importance of testing

The modern world is becoming increasingly dependent on electronic systems in general, and on digital systems such as computers in particular. To be effective, however, such systems must work correctly and, when they fail, techniques must be available to detect the presence of a failure and locate its cause. Many books have been written on techniques designed to produce digital circuits and systems that perform to a given specification, but there are very few books that discuss the related problems of fault detection and fault location—that is, of testing. Traditionally, testing has been relegated to a position of secondary importance relative to design. It is now recognized that the ability to perform correct fault diagnosis is of paramount importance from the point of view of the total cost of design, production, and field maintenance of a digital system. It is also recognized that many of the requirements of testing should be implemented at the design stage.

Despite these comments, there is still considerable ignorance as to what testing requirements are and how they might influence the design of a digital circuit. This book is an attempt to dispel some of this ignorance by describing in some detail the processes involved in testing a digital printed-circuit-board in order to diagnose a fault condition. Inevitably, these processes can be made easy or difficult according to the degree of recognition of testing requirements afforded by the logic designers and by those who transfer a design to a printed-circuit-board. This, the important subject of testable design, is also covered in the book.

About the book

It should be emphasized that this book is an introduction to testing digital boards. It is not intended as an exhaustive treatment of the subject, neither is it intended to be too formal. Indeed, the grammatical style is deliberately informal and the technical emphasis is on the practical aspects of the subject. In this respect therefore, there are certain omissions; these are dealt with below.

The book itself contains 10 chapters and an appendix. Chapter 1 is an introduction to the terminology of testing and, by means of a simple example, sets the scene for the remainder of the book.

Chapters 2 and 3 introduce the subject of test-pattern generation by means of examples designed to illustrate particular aspects of the subject. Chapter 2 is restricted to logic circuits that do not contain stored-state devices or global feedback—that is, to combinational circuits—whereas Chapter 3 looks at the additional problems created by these circuit features. Both these chapters contain a number of exercises together with worked solutions.

The important subject of measuring the performance of a set of tests is discussed in Chapter 4. The evaluation is in terms of the percentage of faults covered from a target fault list and is known as the fault cover of the tests. Most of the chapter describes the structure and use of a fault simulator to produce fault-cover information.

So far, the book has concentrated on the fault-detection aspects of the test patterns. Another major consideration in testing is the degree of fault resolution available from the results of the tests should the board fail. This is the subject of Chapter 5, which introduces the notion of pin-state testing and signature testing. The use of a simulator-generated fault dictionary is also described.

Chapter 6 considers the range of testers available to assist the production and field maintenance of printed-circuit-boards. The range is from device testers to bare-board testers to loaded board (functional) testers.

The chapter concentrates on various attributes of functional board testers. It concludes with comment on the effect on board yield of imperfect device yield and fault cover and also discusses the life-cycle costs of testing.

Following Chapters 1–6, we summarize the main stages of producing a documented test program, ready for use in a production or field-servicing environment. These stages are described in Chapter 7.

Chapter 8 considers the extra complications to testing caused by bus-structured designs utilizing complex modern devices such as micro-processors. There are considerable problems associated with testing such boards. The chapter lists these problems and also describes how some of them are solved.

The last technical chapter, Chapter 9, looks at the whole area of testable design. Chapter 9 starts by describing two design strategies that have met with some success in industry. The first is called Scan-In, Scan-Out; the second is called Signature Analysis. The chapter concludes with a series of practical guidelines for design that seek to ease the problems of generating the tests and to ensure correct and unambiguous fault diagnosis.

The final chapter of the book contains a comprehensive guide to the various sources of further information on testing. In addition to listing current textbooks and conferences on testing, Chapter 10 describes other methods of retrieving technical data, such as the INSPEC service, the Science Citation Index, and the NTIS and STAR sources. The chapter concludes with a selected bibliography of recent papers on specific topics discussed in the earlier chapters of the book.

Finally, the book concludes with an appendix on the D-algorithm. This is the best-known algorithm for generating test patterns for specific fault conditions and, although its application to sequential circuits is limited, the algorithm is significant for the concepts embodied in it. An important derivative of the D-algorithm is the LASAR algorithm; this algorithm is also described in the appendix.

Omissions from the book

There are a number of omissions from the book.

First, the generation of tests by the method of Boolean difference has not been described. This technique is based on a derivation of the fault-free algebraic Boolean function of a logic circuit. The function is then modified according to how the circuit behaves in the presence of a given fault condition. The modified expression is algebraically exclusively-ORed with the fault-free version to produce the input conditions to allow differentiation between the two functions, that is, to produce a test, or set of tests, for the given fault. This approach has received considerable attention from the theoreticians (including the author) but suffers in practice from the need to generate and manipulate the Boolean expressions. The method is also limited to combinational circuits and, for these reasons, has found very little practical application.

Another omission from the test-pattern generation side of the book is the state-table analysis technique. This technique produces what is known as a "checking experiment" for verifying the behavior of a sequential circuit. Unfortunately the method is based on an analysis of the state-sequence flow as defined by the state-diagram or state-table and, in practice, such information rarely exists and is difficult to generate for a realistic sequential circuit.

Finally, there is very little discussion in the book on providing tests for truly random multiple faults or for intermittent faults. Faults are considered to occur singly or, if multiple, are considered to be unidirectional (all stuck low or all stuck high but not both low and high simultaneously) and caused by a bridging failure mechanism. In practice, bridging faults are one of the more significant failure mechanisms on a printed-circuit-board, particularly during the construction phase, and the test programmer certainly needs to guard against such faults. The general problem of providing cover against all possible occurrences of multiple faults is not practical however, and the experience of industry is that most multiple faults are trapped, albeit one at a time, by a set of tests designed to catch single faults and select bridging faults only.

Similarly with intermittent faults—the best solution here is to reduce the possibility of intermittent faults by correct design practices. Unfortunately, this is not always possible, and intermittent faults do occur and are created by a variety of failure mechanisms—bad contact between mating connectors, component ageing, etc. From a testing point of view, there is very little that can be done except to wait for the fault to become permanent (or go away), or to attempt induction to a more permanent form by localized temperature cycling.

Who should read the book?

There are four groups of people who should read at least parts of this book: electronic engineers (or others) about to embark on a career as a test programmer; practicing logic designers; managers of test programmers and test systems; and university and polytechnic students who are studying digital electronic engineering.

Most of the book is aimed at those about to become involved in preparing and evaluating test programs for digital boards. Such people should read the whole book. Managers of test programmers and test systems should read at least Chapters 1, 6, and 7 to gain an awareness of test programming activities and how they might be scheduled, and of the testers themselves. Practicing logic designers are strongly urged to read Chapter 9 on testable design practice. Many complex and therefore costly testing problems could be avoided if the designer had had a better appreciation of testing; Chapter 9 is designed to provide this appreciation. (There is some overlap between the contents of Chapter 9 and other chapters in the book. This is caused by the desire to make Chapter 9 relatively self-contained so that it can be read with minimum reference to the other chapters.)

Finally, the book should serve as useful background reading for students of digital engineering at universities and polytechnics. If the subject features prominently in an individual syllabus, then the book could become a course text.

An essential requirement for all categories of reader is that they have some knowledge of electronic engineering together with a familiarity with common logic devices and with the principles of logic circuit design.

Acknowledgments

This book is the result of some ten years' involvement in the subject of testing, both at an undergraduate and postgraduate teaching level, and in research. I am indebted to numerous students who have, by their questions, forced me to consolidate my understanding of the subject.

Recently however, I returned to industry and had the dubious pleasure of trying to practice what I had been preaching. Much of the material in the book, although based on the earlier university courses, has been considerably tempered by this practical experience. I am also therefore indebted to my colleagues at Cirrus for tolerating my questions and for contributing their own extensive practical experience.

Specifically, I would like to acknowledge the assistance of the following people: Gordon Robinson and John Ibbotson, both of Cirrus, for "volunteering" to proofread parts of the manuscript; Clive Crossley, Chairman of Cirrus, for his constant support and encouragement; Colin Maunder of the British Post Office for his very careful and constructive reading of the whole manuscript; Doug Lewin for pointing me in the right direction ten years ago and for waiting so patiently for the manuscript; Sandra Farrow and Jean Staley of Cirrus for converting my handwriting into word-processor data files and for tolerating my incessant requests for edits to the text; and last, but not least, my wife Carol for extensive moral support backed up by considerable typing effort.

R. G. Bennetts
September, 1980

Chapter 1
Introduction to Testing

This chapter introduces the basic concepts and requirements of digital board testing and the contents of the remaining chapters.

Throughout the book, the logic symbols used to depict gates, flip-flops, and other digital devices will be as described in the American IEEE standard 91-1973. This standard is summarized in Figure 1.1.

1.1 Objectives and Terminology

The primary objective of testing digital printed-circuit-boards (PCBs) is to detect the presence of hardware failures induced by faults in the IC fabrication processes, by PCB manufacture, by assembly processes, or operating stress or wearout mechanisms. Such testing is often referred to as "go/no go" testing. The secondary objective is to locate the cause of a fault with enough precision and correctness to enable repair to be carried out. This form of testing is referred to as "diagnostic", the word diagnosis implying both detection and location. The word "fault" relates to the physical failure mechanism, e.g., a solder splash, whereas the term "fault-effect" relates to the logical effect of the fault on a signal-carrying node, e.g., the node is stuck-at logic 0.

The word "error" refers to the condition of the circuit when it contains a fault. "Failure" of the circuit is therefore the result of the

1

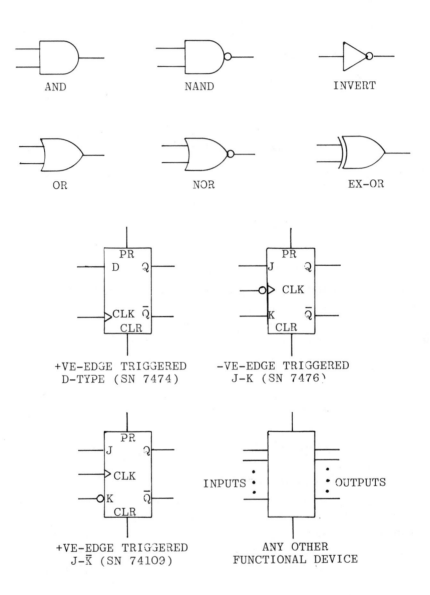

FIGURE 1.1 Logic Symbols (IEEE Std. 91-1973)

circuit being in an error condition, the circumstances being such that the fault causes the circuit to perform incorrectly.

The difference between "error" and "failure" can be illustrated by considering an automobile carrying a spare tire which, unknown to the driver, is flat because of a faulty valve. The driver may drive many miles before the need arises to change a wheel. Until that time, the car is in an error condition but has not yet failed.

In general a digital PCB is an assembly of basic logic gates, flip-flops, and more complex digital devices such as shift registers, counters, ROMS, RAMS, and even microprocessors. It may also contain other active devices such as transistors and operational amplifiers; discrete passive components such as pull-up resistors and decoupling capacitors; and structural features such as feedback and buses. Normally, however, the board can only be driven (stimulated) by a tester from certain positions along its edge connector. Similarly, it can usually only be sensed (monitored) at other positions along the edge connector.* Inputs that can be driven and outputs that can be sensed are referred to as "primary inputs" and "primary outputs" respectively. In this connection, the terms "test," "test pattern," or "test input/output vector" mean a specified primary input stimulus plus a known fault-free primary output response. As noted earlier, the word "fault" means any failure mechanism or malfunction that alters the logical behavior of the circuit; the term "fault-cover" refers to the total set of faults detected either by a single test or by a set of tests. In the latter case, the term "fault-cover" refers to the cumulative cover of each of the single tests that constitute the full set of tests.

1.2 Faults and Fault-effect Models

When modelling the behavior of a logic circuit in the presence of a fault, it is often convenient to isolate the actual failure mechanism from its effect. A popular fault-effect model is the "stuck-at" model, that is, a logic node is assumed to be stuck at one of its two logic levels, stuck-at-one or stuck-at-zero. Figures 1.2, 1.3, and 1.4 illustrate how three different types of failure mechanism can create stuck-at fault effects.

*Further access can be gained via a set of "bed-of-nail" probes—see Chapter 6, section 6.1.2.

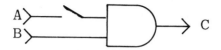

FAULT 1   TTL AND gate input open-circuit

The effect of this fault is to make
input A appear to be stuck-at-1 (an
open-circuit TTL connection floats at
about 5v).

FIGURE 1.2 Open-Circuit Fault

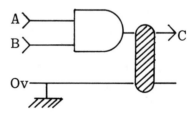

FAULT 2   Short-circuit to Ov on gate output

The effect of this fault is to make
output C stuck-at-0.

FIGURE 1.3 Short-Circuit Fault

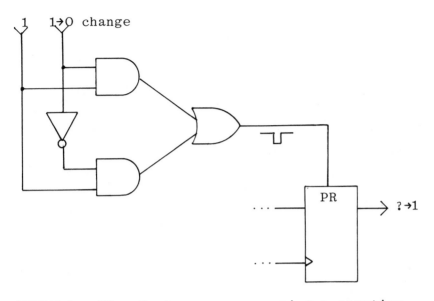

FAULT 3   Signal-change race causing a negative
           glitch (design fault)

This fault is more devious to find. The
negative glitch on the preset input to the
flip-flop may cause the flip-flop to preset
to 1. If it was originally 1, then this is
of no consequence. If originally zero,
however, the flip-flop output changes
incorrectly and appears to be stuck-at-1 (but
not permanently so).

FIGURE 1.4 Hazard Condition

The stuck-at fault-effect model is popular as a basis for test generation, but is not guaranteed to cover all causes of failure. Figure 1.5
summarizes the main causes and effects of failures in PCBs containing
SSI or MSI devices.

Recently, other models have been suggested to supplement the
stuck-at model. Of particular value is the "bridging fault" model and,
more recently, the "propagation delay" model. The bridging fault
model is useful for modelling track-bridging faults that result in a

FAULT TYPE	CAUSE OF FAULT	ELIMINATED BY	EFFECT OF THOSE REMAINING
LAYOUT	CROSSTALK, LINE REFLECTION, GROUND PLANE INCOMPATABILITIES, POWER RAIL NOISE, VIOLATION OF FAN-IN OR FAN-OUT RULES, EXTERNAL ELECTRICAL NOISE.	CORRECT APPLICATION OF LAYOUT RULES	SOLID AND INTER-MITTENT STUCK-AT FAULTS
CONSTRUCTION	INCORRECT INTERCONNECTIONS, INCORRECT IC PACK-AGES, SOLDER SPLASH, TRACK BRIDGING, IC PIN BRIDGING, DRY OR UNSOLDERED JOINTS.	CAREFUL CONSTRUCT-ION PROCESSES AND VISUAL INSPECTION	MOSTLY COVERED BY STUCK-AT MODEL EXCEPT SOME BRIDGING FAULTS
IC INTERNAL FAILURE	MASK, PHOTO-DEPENDENT, SILICON METAL OR CHIP-TO-PACKAGE (OPEN-CIRCUIT/SHORT-CIRCUIT) DEFECTS	CAREFUL CONSTRUCT-ION, SCREENING AND SELECTION PROCESSES	MOSTLY COVERED BY STUCK-AT MODEL EXCEPT CERTAIN METAL-TO-METAL SHORTS
ENVIRONMENT	COMPONENT DEGRADATION DUE TO HIGH HUMIDITY OR THERMAL CONDITIONS OR ELECTRICAL NOISE	USE OF COMPONENTS WITH CORRECT ENVIRONMENTAL SPECIFICATION	COVERED BY STUCK-AT MODEL BUT MAY BE INTERMITTENT
TIME-DEPENDENT	COMPONENT AGEING FAULTS (ALUMINIUM MIGRATION IN IC'S, RESISTOR OR CAPACITOR DEGRADATION), MODIFICATIONS!	PREVENTATIVE MAINTENANCE	COVERED BY STUCK-AT MODEL
DESIGN AND IMPLEMENTATION	CRITICAL RACES, STATIC, DYNAMIC AND ESSENTIAL HAZARDS	CORRECT DESIGN METHODS AND VALIDATION	SOME COVERED BY STUCK-AT MODEL BUT MAY BE INTERMITTENT

SOLID STUCK-AT FAULTS CAN BE EITHER SINGLE OR MULTIPLE. IF MULTIPLE, THE POLARITY OF THE STUCK-AT NODES IS USUALLY THE SAME (ALL STUCK-AT-O OR ALL STUCK-AT-1). FAULTS SUCH AS THESE ARE CALLED UNIDIRECTIONAL MULTIPLE FAULTS OR, MORE COMMONLY, BRIDGING FAULTS.

FIGURE 1.5 Causes and Effect of PCB Failures (SSI/MSI Devices)

wired-OR or wired-AND effect and is discussed in more detail in Chapter 2. The propagation delay model is used to model the effect of failures due to timing defects—either degradation of device propagation delays or out-of-tolerance timing relationships between various devices.

1.3 Testing Activities and Requirements

The first step in producing a test program is to generate the test patterns. This is the "test-pattern generation" (TPG) activity. The requirements are as follows:

    (i) a logic diagram for the board-under-test (B-U-T);
    (ii) device data sheets;
    (iii) a fault-cover target;
    (iv) a methodology for producing the tests.

Generally, the fault-cover target is expressed in terms of the fault-effect models, e.g., a test for each circuit node either stuck-at-1 or stuck-at-0.

Various methods for producing the tests are described in Chapters 2 and 3.

The second activity is to evaluate the effectiveness of the test patterns. This is the "test-pattern evaluation" (TPE) activity. The evaluation of the test patterns is carried out against the fault-cover target and requires either a known-good-board (KGB) on which faults can be physically inserted to measure the performance of the test programs, or the availability of a logic simulator to simulate the behavior of the board for each of the test input patterns in order to produce fault-free reference data for the output response. The simulation is then repeated for one of a given set of faults to assess whether the test pattern is capable of detecting that particular fault. Detection is achieved if the value of at least one primary output differs according to the presence or absence of the fault. This process of assessment is repeated for all the faults listed in the fault-target list.

The use of a logic simulator is discussed in more detail in Chapter 4.

The third and last activity is to apply the tests to a real board containing a fault. This is referred to as the "test-application and fault-finding" or TAFF activity. To do this requires a tester plus documentation to support the use of the test program. Techniques for ensuring correct diagnosis are discussed in Chapter 5.

To set the scene for the next four chapters, each of these activities will be considered and related to a shift-register example.

1.3.1 Test-Pattern-Generation Strategies

There are two fundamental strategies for writing test patterns for logic boards. The first is called the "functional" or "behavioral" strategy, and seeks to provide a sequence of input changes with corresponding output responses that will verify the function of the PCB. The second strategy is called the "structural" or "fault-oriented" strategy, and is based on providing tests to detect the presence of a hardware fault taken from a set of predefined faults or, more usually, a set of predefined fault-effects. Figure 1.6 illustrates the basic difference between these two strategies.

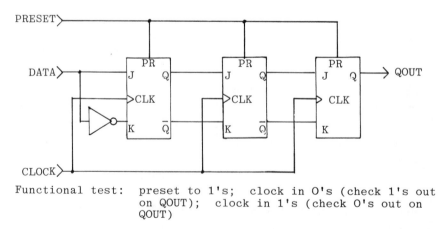

Functional test: preset to 1's;   clock in 0's (check 1's out on QOUT);   clock in 1's (check 0's out on QOUT)

Structural test: write a test for each circuit connection (node) either stuck-at-1 or stuck-at-0.

FIGURE 1.6 3-Stage Shift Register

For this circuit, the functional test strategy would seek to provide inputs in order to demonstrate the shift-register nature of the circuit. One way to do this is to preset to ones, clock in zeroes, checking for ones on QOUT, and then to clock in ones, checking for zeroes on QOUT, and finally to preset back to ones again. The alternative structural strategy would be based on a list of faults to be covered. This list, derived from the fault-cover target, is called the "fault-list." For the purpose of our discussion, we will assume a fault-list of each node stuck-at-1 and stuck-at-0. Each test, therefore, is designed to detect the presence of a particular fault-effect on a particular node. Taken collectively, the final set of test patterns should cover all the original faults in the fault list. Methods for generating such tests are covered in the following two chapters. Note that the first set of tests, based on the functional strategy, is not primarily concerned with the possible fault conditions that might exist within the circuit, whereas the second set of tests, based on the structural strategy, is not primarily concerned with the overall function of the circuit. In practice, both sets of tests will be evaluated against their ability to detect actual faults, and in this respect structural testing is closer to the main objective of testing. Functional testing is generally easier, however, and usually produces valid test patterns anyway. Further discussion on the marriage of these two strategies to produce test patterns is contained in Chapter 3.

1.3.2. Test-Pattern Evaluation

The second activity of test programming is to evaluate the effectiveness of the test patterns. Test-pattern evaluation can be carried out either by means of a logic simulator or by inserting real faults on to a known good version of the board. Evaluation by simulation is a popular but costly solution. The basic strategy is to exercise a model of the circuit and assess its behavior under application of the test patterns in order to produce fault-free behavioral reference data. The simulation is then repeated with a particular fault taken from the fault list and imposed on a node within the circuit model. A comparison is then made between the response in the presence of the fault and the fault-free reference values. Any difference in the two sets of output values means that the fault is detectable. This process is repeated for each fault in the list of faults to be covered. To be of any value, however, the simulator must model the behavior of the circuit accurately. This means including a time parameter to allow modelling of real timing relationships, such as propagation delay and edge rise and fall effects. It is this requirement for accuracy that makes both the development and running of a simulator a costly exercise.

The alternative cheaper strategy is to insert real faults on to a known good copy of the board and to observe, on the tester itself, whether the fault is detected. This strategy is relatively easy but suffers from two main limitations. The first is that the types of fault that can be inserted are limited to those that do not cause irreversible physical changes, such as track cutting, to the board, or that do not cause irreversible damage to the devices on the board. This limitation does not apply to simulated models of the board.

The second limitation of fault insertion is that it requires a known good copy of the board. This may not be a problem if the fault insertion is carried out on the board used for developing the test program. By the time the program is ready for evaluation the behavior of the board has usually been verified against the logic diagram and confidence is high that it is indeed a known good board.

1.3.3. Test-Application and Fault-Finding

In this section, the third and last activity of test programming—that of test-application and fault-finding—is examined. For this some understanding is required of what a tester contains and how it operates. Figure 1.7 shows the basic architecture of a tester.

The figure shows a number of features. The first is the interface between the tester and the board-under-test. This interface is at the "driver/sensor" pins of the tester. As the name suggests, these are pins on the tester that can be used either to drive information on to the board or to sense the response of the board. Generally, the physical position of each pin is fixed, although not necessarily its status as a driver or sensor. Often, however, at least one pin is attached to the end of a coaxial cable and is used as a roving sensor probe. This probe is used to locate faulty devices on the board. The control of the driver/sensor pins, including the guided probe, is carried out by a main processor which is usually a

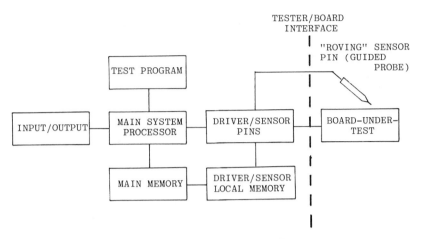

FIGURE 1.7 Architecture of a Basic Tester

minicomputer or microcomputer. This computer would have the normal keyboard and line-printer input-output facilities, together with back-up memory to hold a library of test programs for different board types. It is also usual for the driver/sensor pins to have their own local memory to speed up the test application rate. The main functions of the tester, therefore, are as follows:

(i) to retrieve, interpret, and apply test data;
(ii) to monitor the response of the board and indicate the pass/fail status;
(iii) to determine the likely cause of failure in the event of the board not passing the test.

Chapter 5 discusses fault location either by guided probe or by means of a fault dictionary created by a logic simulator. Chapter 6 deals with the various types of tester in terms of their methods of operation and architectural features.

1.4 Structure of a Test Program

To complete this chapter, another simple circuit will be considered and the construction of a suitable test program outlined. The circuit is shown in Figure 1.8.

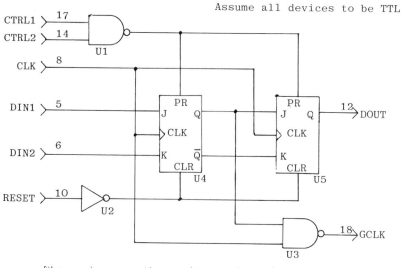

Assume all devices to be TTL

The numbers on the primary input/output lines are the absolute numbers of the driver/sensor pins which connect with these lines when the board is loaded into the tester.

FIGURE 1.8 Simple Example Circuit

This circuit contains two JK flip-flops connected in a shift-register configuration. The data inputs to the first stage of the shift-register are called DIN1 and DIN2 respectively and the data output of the second stage of the shift-register is called DOUT. Both stages of the shift-register are clocked synchronously by means of the same clock coming from the input terminal marked CLK. Both stages also have a common RESET facility connected to the clear line of each flip-flop, and a common preset facility derived from the two inputs CTRL1 and CTRL2. Finally, the clock and the output of the first stage of the shift register are gated together to provide a gated clock, called GCLK.

1.4.1 Test Strategy for Fig. 1.8

A possible test strategy, based on the functional approach, for this particular circuit could be as follows: first, initialize the two flip-flops using the RESET input which is common to both the clear lines. To a certain extent, this can be checked by looking at the value of the DOUT and GCLK lines. Once the circuit is initialized the reset line should be released—that is, put to its enable rather than inhibit level—and the clocked behavior of the shift-register tested by changing the data on DIN1 and DIN2 and clocking accordingly. When the operator is satisfied that the shift register is capable of accepting both levels of data, logic one and logic zero, and also of propagating this data under control of the clock pulse, the action of the preset control facility can be tested. To do this requires changes to the values of CTRL1 and CTRL2 so as to cause both flip-flops to preset. The effect of these changes can be observed on the DOUT and GCLK lines.

As a last test the circuit should be returned to a known state by reinitializing using the RESET facility. Figure 1.9 shows the timing diagram associated with this testing strategy.

The initialization, clock, and preset tests are shown in Figure 1.9.

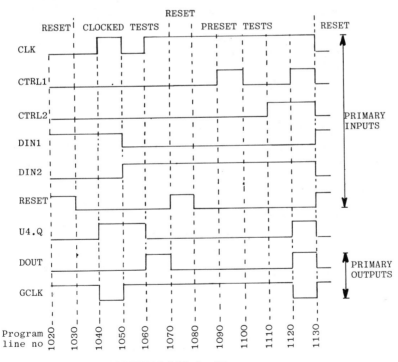

FIGURE 1.9 Timing Diagram

Note that there is also a test to show that both CTRL1 and CTRL2 should necessarily be high before the preset occurs. If we were now to move a ruler vertically across this timing diagram, noting down the levels of the primary inputs together with the expected levels of the primary outputs, then in effect we would be writing down the tests for this circuit. These tests are shown in Figure 1.10.

Note the format of the test program. The symbolic names CLK, CTRL1, CTRL2, DIN1, DIN2, and RESET have been defined to be INPUTS—that is, inputs that the tester can stimulate. Similarly, DOUT and GCLK have been defined to be OUTPUTS—that is, outputs that the tester can monitor. Each test has now been defined by specifying the values of these inputs and outputs as either high or low. IL CLK means that the clock input is to be driven low. Similarly, IH RESET means that RESET is driven high. OL DOUT specifies the value that is expected on the DOUT output as a result of a particular input stimulus. Finally, the "X" statement that occurs at the end of every line should be interpreted as an instruction to the tester to carry out the test, i.e., to test the status of

```
1000  INPUT CLK,CTRL1,CTRL2,DIN1,DIN2,RESET
1010  OUTPUT DOUT,GCLK
1020  IL CLK,CTRL1,CTRL2,DIN2;IH DIN1,RESET;OL DOUT;OH GCLK;X
1030  IL RESET;X
1040  IH CLK;OL GCLK;X
1050  IL CLK,DIN1;IH DIN2;OH GCLK;X
1060  IH CLK;OH DOUT;X
1070  IH RESET;OL DOUT;X
1080  IL RESET;X
1090  IH CTRL1;X
1100  IL CTRL1;X
1110  IH CTRL2;X
1120  IH CTRL1;OH DOUT;OL GCLK;X
1130  IL CLK,CTRL1,CTRL2,DIN2;IH DIN1,RESET;OL DOUT;OH GCLK;X
```

```
IL MEANS INPUT LOW
IH MEANS INPUT HIGH
OL MEANS OUTPUT LOW
OH MEANS OUTPUT HIGH
```

FIGURE 1.10 Test Program for Figure 1.9

the inputs and the outputs as they exist on the board and as they should be according to the specification in the program.*

1.4.2 Completing the Program

There are other requirements, which are either tester-specific or board-specific, needed to complete the test program shown in Figure 1.10. Briefly, these are such items as the following:

(i) Program identification data (test programmer, date, board number, revision level, etc).

(ii) Specification of the relationship between the symbolic names used in the logic diagram and program and the absolute numbers of the driver/sensor pins to which they are connected. (This relationship is shown in Figure 1.8.)

(iii) Specification of the driver high, driver low, and sensor threshold voltages. For example, for a TTL board, driver high would be +5v, driver low 0v, and sensor threshold 1.4v. An IH CLK statement would therefore result in a +5v signal being applied to driver/sensor pin 8. Similarly, a test OL DOUT would only be passed provided the sensed value on pin 12 was below the value of 1.4v. (Normally TTL logic 0 lies between 0v and 0.8v, and logic 1 between 2.4v and 5v. A single threshold of 1.4v is therefore acceptable for both values.)

(iv) Setting the programmable power supply pins on the tester to levels appropriate to the board's requirements, e.g., +5v supply and 0v ground return for TTL devices. This instruction is usually followed by an appropriate delay to allow the power supply voltage to rise to its programmed level.

(v) Setting any important tester control values, such as the time delay between the application of the input stimuli (on the driver pins) and the test on the status of the output response (on the sensor pins). In general, the tester should wait for a period of time at least greater than the worst-case propagation delay of

---

*The format of the statements used in this test program conforms to that used by the GenRad 2225 Portable Service Tester. The language for this tester is based on BASIC.

the board before testing the status of the board's response. This time is called the "strobe" time.

(vi) Setting the "skew" or "broadside" option for input pin changes. Consider a state such as:

IL CLK, CTRL1, CTRL2, DIN2

These changes can either take place one at a time and in the order specified or they can all happen simultaneously. The former mode is called "skew", the latter "broadside." Usually, the skew mode is preferred but, for a parallel bus for example, the broadside mode may be required.

Incorporating these extra facilities into the program produces the final version shown in Figure 1.11.

```
10 REM TEST PROGRAM FOR FIG 1.8
20 REM AUTHOR R.G.BENNETTS, JUNE, 1980
30 PRINT "TEST PROGRAM FOR FIG. 1.8"
40 EQUATE (CTRL1,17),(CTRL2,14),(CLK,8),(DIN1,5)
50 EQUATE (DIN2,6),(RESET,10),(DOUT,12),(GCLK,18)
60 SET (1,0),(2,5),(3,1.4); REM D/S REF VALUES
70 SET (10,5); REM POWER SUPPLY 10 TO +5V
80 DELAY 50000;REM 50 MILLISECONDS DELAY
90 STROBE 10;REM 10 MICROSECONDS STROBE TIME
100 SKEW
1000 INPUT CLK,CTRL1,CTRL2,DIN1,DIN2,RESET
1010 OUTPUT DOUT,GCLK
1020 IL CLK,CTRL1,CTRL2,DIN2;IH DIN1,RESET;OL DOUT;OH GCLK;X
1030 IL RESET;X
1040 IH CLK;OL GCLK;X
1050 IL CLK,DIN1;IH DIN2;OH GCLK;X
1060 IH CLK;OH DOUT;X
1070 IH RESET;OL DOUT;X
1080 IL RESET;X
1090 IH CTRL1;X
1100 IL CTRL1;X
1110 IH CTRL2;X
1120 IH CTRL1;OH DOUT;OL GCLK;X
1130 IL CLK,CTRL1,CTRL2,DIN2;IH DIN1,RESET;OL DOUT;OH GCLK;X
1140 END
```

FIGURE 1.11 Complete Test Program

There are still other activities required before this test program can be regarded as fully complete (such as, does it find faults?), but further discussion will be deferred until Chapter 7, in which the ten main stages of producing a test program will be summarized.

1.5 Boards Containing LSI and VLSI Devices

There are certain specialized problems associated with the writing of test programs for bus-structured boards containing LSI and VLSI devices. These problems cannot be appreciated, however, until a basic understanding has been gained of the procedures for SSI and MSI boards. Discussion on this subject is therefore relegated to Chapter 8.

1.6 Testable Design Techniques

The important subject of testability and how it can be incorporated into a circuit, preferably at the design stage, is discussed in Chapter 9. A number of very practical guidelines are there described which will simplify many of the real problems encountered in writing and using test programs.

Chapter 9 also discusses the principles and application of two techniques for enhancing testability—one at the initial circuit design stage and the other at the board design stage. The two techniques are known generically as Scan-In, Scan-Out and Signature Analysis. Both have found application in industry.

1.7 Bibliography

Following Chapter 9, there is a general guide to the sources of more information, followed by a bibliography organized in terms of subject classifications. The reader who wishes to pursue a specific topic will find the textbooks and papers listed a useful starting point.

Chapter 2
Test-Pattern Generation: Combinational Circuits

Chapter 1 introduced the three activities of test programming, namely, test-pattern generation, test-pattern evaluation, and test-application and fault-finding. We return now to the first of these activities and, by way of a series of examples, we will build up our understanding of what is involved in choosing a set of test patterns. In this chapter, the examples will be restricted to simple circuits, without feedback, but in Chapter 3 we will consider circuits containing stored-state devices, such as flip-flops, as well as circuits containing global feedback from one device to another.

In general, each example is designed to illustrate one aspect of the TPG problem. In practice, of course, a test programmer has to consider many different factors before deciding upon a strategy and producing a test program.

Also in Chapters 2 and 3 will be found a number of exercises together with worked solutions. In general, these exercises are designed to consolidate understanding, and should be considered an integral part of the text. The worked solutions are provided for purposes of comparison. The reader should be warned, however, that in a given case there is often no unique solution.

Finally, in the examples that follow, the logic variables (primary inputs and primary outputs) are identified separately from the nodes in the circuit. This is done to avoid confusion between the specified test values on inputs and outputs and the fault-free or faulty nodal values. In general, primary inputs are denoted by subscripted x symbols, e.g., x_1, x_2; primary outputs by subscripted z symbols, e.g., z_1, z_2; and circuit nodes by subscripted c (for connection) symbols, e.g., c_1, c_2.

2.1 Example 1: Functional vs. Structural Approach

The first example, shown in Figure 2.1, is the simplest of all types of logic circuitry and is designed to show the fundamental difference between a functional test set and a structural test set.

The circuit consists of a 2-input NAND gate, and the functional test set is defined by the full truth table for this type of gate, namely:

Test input		Expected output
x_1	x_2	z_1
0	0	1
0	1	1
1	0	1
1	1	0

Note that such a test set is exhaustive and must therefore cover all possible failure mechanisms—either internal or external—that would cause the gate to function no longer as an NAND gate. This circuit is an example of a "combinational" or "acyclic" circuit; that is, there are no stored-state devices, such as flip-flops, nor is there global feedback from one device to another. For the 2-input circuit shown in the figure,

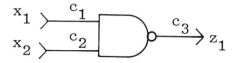

FIGURE 2.1 Example 1

the full truth-table contains 2^2 rows—that is, 4 tests. In general, if the circuit is combinational and contains n primary inputs, then the full functional test would consist of 2^n tests. Unfortunately, 2^n grows very quickly as n increases and this factor tends to limit the practical application of functional testing. The alternative is to derive test-patterns that seek to demonstrate the presence or absence of a class of failure mechanism or, more usually, of a class of fault that have a predictable effect on the logical behavior of the circuit. By far the most popular of these fault-effects is the stuck-at-1, stuck-at-0 model, discussed in Chapter 1.

This model is used as a basis for test-pattern generation in the following way. First, it is assumed that every real failure mechanism will manifest itself by causing at least one of the logic-carrying connections in the circuit to appear to be stuck either at logic 1 (s-a-1) or at logic 0 (s-a-0). It is then suggested that if a set of tests can be derived that are guaranteed to prove that no single connection is either s-a-1 or s-a-0, then the circuit is considered fault-free. The starting poing therefore is to write down a list of s-a-1 and s-a-0 faults to be covered by the set of test patterns. This is the "fault list." One of these faults is then selected as the possible fault and a suitable test-pattern derived by a process that will be explained, conceptually at least, in the following example. Once this test has been derived, the next step is to determine what other faults, so far uncovered, would also be detected by this test. This procedure not only helps to shorten the list but also prevents the derivation of unnecessary tests.

The process is then repeated starting with another fault from the list and, eventually, tests are generated for all listed faults and the process terminates. The method is summarized below. For this example, it produces three tests instead of the four required for the full functional test.

Step 1   Initial fault list.* $c_1/1$, $c_1/0$, $c_2/1$, $c_2/0$, $c_3/1$, $c_3/0$

Step 2   Generate test for $c_1/1$ Result: $x_1 = 0$, $x_2 = 1$, $z_1 = 1$

Step 3   Any other faults covered by this test?   Result: $c_3/0$

Step 4   Revise fault list.   Result: $c_1/0$, $c_2/1$, $c_2/0$, $c_3/1$

*$c_1/1$ means c_1 s-a-1, etc.

Step 5   Generate test for $c_1/0$ Result: $x_1 = 1, x_2 = 1, z_1 = 0$

Step 6   Any other faults covered by this test?   Result: $c_2/0, c_3/1$

Step 7   Revise fault list.   Result: $c_2/1$

Step 8   Generate test for $c_2/1$ Result: $x_1 = 1, x_2 = 0, z_1 = 1$

Step 9   Any other faults covered by this test?   Result: $c_3/0$ (already covered)

Step 10   Revise fault list.   Result: Empty (all faults covered)

Step 11   Final test set and fault-cover:

Test Input		Expected Output	Faults covered on		
x_1	x_2	z_1	c_1	c_2	c_3
0	1	1	s-a-1		s-a-0
1	1	0	s-a-0	s-a-0	s-a-1
1	0	1		s-a-1	s-a-0

Note that although these tests would detect any node s-a-1 or s-a-0, they would not detect the fact that the actual gate inserted was an Exclusive-OR gate rather than a NAND gate.

In practice, the structural technique is not so easy to apply as this example suggests; later examples will show various reasons why this is so. What can already be observed is that the technique is based on the assumption that if it can be shown that the circuit contains no connection either stuck-at-1 or stuck-at-0, then the circuit is fault-free. The validity of this assumption is sometimes questionable for certain types of failure mechanism and, in writing structural test patterns, the test programmer should not be bound by the stuck-at model. It is sometimes very necessary to write specific tests for certain types of failure that are known to occur, and whose logical effect is more complex than just to cause a connection to become stuck at one of the logic levels. A good example of this is the ''bridging fault.'' A bridging fault is created when two or more independent but adjacent connections become inadvertently joined together—by a solder splash, for example.

The causes of bridging faults are numerous and are listed below:

(a)   Inside the chip: cross-over connection as a result of defective masking, etching deficiencies, aluminum migration, insulation breakdown, surface or oxide defects, channeling, pinholes, etc.

(b) At PCB level: solder splash, unetched copper, excess, bare, or loose cross-over wires, excess solder, etc.

(c) At backplane level: defective wire-wrapping, excess solder or solder splash on printed-circuit mother boards, bent wire-wrap pins, defective wire insulation, etc.

Most logic families exhibit a wired-logic property if gate leads are connected together. This property is used to model bridging faults. The faults are referred to as AND-type or OR-type bridging faults depending on the particular wired-logic result.

Such faults are not necessarily detected by a set of tests designed to detect all single s-a-1, s-a-0 faults. Consider the simple circuit shown in Figure 2.2, for example.

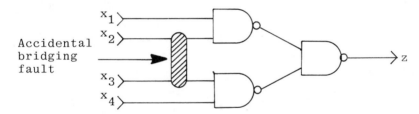

FIGURE 2.2 Bridging Fault Example

This circuit is fully tested for all single s-a-1, s-a-0 faults by the four tests:

$$T_1 = \left\{ \overline{x_1}x_2x_3\overline{x_4}/\overline{z},\ x_1\overline{x_2}\overline{x_3}x_4/\overline{z},\ \overline{x_1}x_2x_3x_4/z,\ x_1x_2x_3\overline{x_4}/z \right\}$$

Unfortunately, the value of x_2 and x_3 is specified as the same in all four tests. This means that an accidental bridging fault between these two inputs will not be detected by the set of tests T_1, irrespective of the wired-OR or wired-AND nature of the bridge. In this case, the problem is overcome by rearranging the first two tests to produce:

$$T_2 = \left\{ \overline{x_1}x_2\overline{x_3}x_4/\overline{z},\ x_1\overline{x_2}x_3\overline{x_4}/\overline{z},\ \overline{x_1}x_2x_3x_4/z,\ x_1x_2x_3\overline{x_4}/z \right\}$$

This sequence will now detect a wired-OR bridging fault but not a wired-AND. Detection of the wired-AND will require an additional test—$x_1\bar{x}_2x_3x_4/z$ or $x_1x_2\bar{x}_3x_4/z$.

It is also possible for a bridging fault to turn a combinational circuit into a circuit that possesses feedback. Detection of the fault then becomes a function not only of the tests themselves but also of the order in which the tests are applied.

For both the functional and the structural test patterns for Example 1, Figure 2.1, no attention was paid to the order in which the tests were applied and, for this example and the faults listed, the order is immaterial. In general, the order becomes important when the circuit contains either stored-state devices or global feedback, or both.

The general conclusion about the relationship between real failure mechanisms and the stuck-at fault-effect model is that the model is a useful and firm basis for generating test patterns for most types of fault. There are some categories of failure, both at device level and between connections on a board, for which the stuck-at model is inadequate, however. A test programmer should develop an awareness of these inadequacies and compensate with additional tests where necessary.

Returning to Example 1, we note that the structural technique succeeded in reducing the number of test patterns by only one test. In practice, the reduction in the final number of test patterns is usually more dramatic than this. For an n-input acyclic circuit requiring 2^n functional tests, the number of structural tests needed to cover all single stuck-at faults is of the order $2n$. This figure represents a significant decrease on 2^n as n becomes large, and it is for this reason that the structural technique has received considerable attention, both from the theoreticians and the practitioners of testing. In the following example, the process of generating a test for a specific stuck-at fault condition will be more closely examined.

2.2 Example 2: Sensitive Path Concept

Example 2 is shown in Figure 2.3.

In this example, we will show how a test might be generated for a stuck-at-1 fault on c_1.

First, note that the fault-free value on c_1 must be opposite to the postulated stuck-at faulty value. It is not possible to detect a stuck-at-1

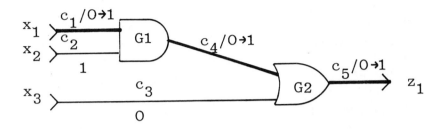

FIGURE 2.3 Example 2

fault condition if the fault-free value is also a 1. This requirement sets x_1 to 0.

Second, we need to set the other input to G1, to what the logic designer would call the "enable" level for the gate. For the AND gate, this level is 1. This ensures that if the postulated fault is present, then the output of G1 will change from its fault-free value of 0 to a value of 1, thereby propagating the information about the presence of the fault. This propagation of fault information is the main objective of structural test-pattern generation and, ultimately, that objective is to drive the information forward to a point where it can be observed by the tester. For Example 2, this means driving through G1 and then through G2 to primary output z_1.

The third stage is to set the c_3 input to the OR gate to the enable level for this type of gate—that is, to 0. If the fault is present, the effect of the fault is propagated first through gate G1 and then through gate G2, eventually appearing at the primary output. The test is defined by the primary input values $x_1 = 0, x_2 = 1, x_3 = 0$, with the fault free output response of $z_1 = 0$. If the stuck-at-1 fault is present on c_1, a change of value will occur at the output of G1 and this, in turn, will cause a change of value at the output of G2. Note that the flow of fault information could have been blocked by selecting the "inhibit" values on either the c_2 input to gate G1 or the c_3 input to gate G2. Setting c_2 to 0 instead of to 1 would have clamped the output of G1 to 0 irrespective of what the value was on the c_1 input. Similarly, setting c_3 to 1 would have blocked the flow of information through G2. This idea of blocking fault-data is just as important as propagating fault-data; the reason will become clear in a later example.

Returning to the figure, we can now formalize the concepts embodied in the generation of this test. The path along which the fault information propagates, starting from the source of the fault, is called a "sensitive" path. The source of the fault is called the "fault-generator." All other connections that go to make up the sensitive path are called "fault-transmitters." The other connections to gates G1 and G2 that are not part of the sensitive path, but that are fixed at their enable levels to allow forward propagation along the path, are said to have "fixed values" on them. This method for generating tests is called the "sensitive-path" technique to distinguish it from other, more algebraic techniques.

To summarize so far: the basic strategy for structural testing is to start with a postulated fault, taken from the fault list, and generate a test pattern. The most popular technique for generating such patterns is the sensitive-path technique. This method seeks to create a path from the source of the postulated fault to at least one of the primary outputs of the circuit. If the fault is present, its effect propagates through to the primary output, changing the expected fault-free value. Creation of the sensitive path requires the assignment of fixed logic values to the other inputs to the logic gates and to other devices along the route. These fixed values must, eventually, be established from other primary inputs to the circuit. This process is called "consistency checking" and can often be quite complex. In the following example, Example 3, we look at these concepts again, and introduce the idea of marrying the functional and structural testing strategies.

2.3 Example 3: Functional Plus Structural Approach

This example, shown in Figure 2.4, is somewhat unrealistic because of the odd mix of gate types. Nevertheless, the problems illustrated by the circuit are very real.

To produce a set of test patterns for this circuit, we could start with a fault list of stuck-at faults on each node and use the sensitive-path technique to generate suitable tests. This would not be an incorrect strategy but may be unnecessarily long. An alternative, and very often more profitable strategy, is to apply some of the functional tests first and see how many of the stuck-at faults in the fault list can be removed

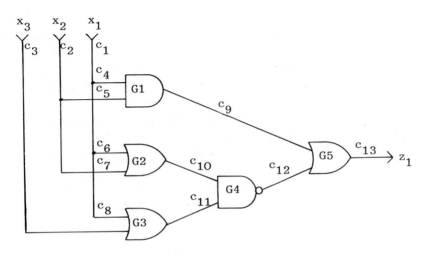

FIGURE 2.4 Example 3

before we are forced to adopt the sensitive-path technique. If a limited functional test can be written down quickly and yet covers 70% – 80%, say, of the faults in the fault-list, then the number of applications of the sensitive-path technique will be considerably reduced.

To illustrate the functional approach, consider the Karnaugh map for the function. The map is shown in Figure 2.5.

Looking at the map, a possible reduced functional test might be to start with the $x_1 = 0, x_2 = 1, x_3 = 0$ values and then move as shown,

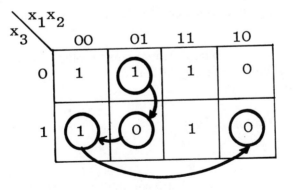

FIGURE 2.5 K-Map

finishing at $x_1 = 1$, $x_2 = 0$, $x_3 = 1$. This procedure would specify four tests and cause the output to change from 1 to 0 to 1 and back to 0 again. The four tests also place both 0's and 1's on each of the primary inputs.

Altogether, therefore, the four tests would seem to be a useful limited set of functional tests. The question now is—how many single s-a-1, s-a-0 faults are covered by these four input changes? Leaving aside for the moment the detail as to how this question is answered, the answer in this case is that all but one fault is covered. The fault that is not covered is a stuck-at-0 fault on node c_9. At this stage, it would be useful to carry out the following exercise based on this example.

Exercise 2.1

Carry out the following exercises on Example 3 (Fig. 2.4).

1. Show that the limited set of four functional tests, T, defined in the K-map of Fig. 2.5, does not detect the presence of a s-a-0 on node c_9 in Figure 2.4.

2. Show that the same set of tests do detect the following faults:
      i) c_1 s-a-0
     ii) c_1 s-a-1
    iii) c_9 s-a-1
    iv) c_{10} s-a-0

3. Using the sensitive-path technique, derive a test for the uncovered fault, node c_9 s-a-0.

For the purpose of simplicity, ignore faults on the individual fan-out branch connections, c_4, c_5, c_6, c_7 and c_8.

Exercise 2.1: Outline Solution

Solution to Part 1

For each of the four tests, the expected fault-free value of c_9 is 0. If the node is stuck-at-0, this is consistent with the fault-free value. Therefore the fault is not detected.

Solution to Part 2

The general procedure is as follows:

Step 1: For each test input, determine the full set of fault-free nodal values.

Step 2: Select the first fault (c_1 s-a-0) and repeat Step 1, but allow the fault to override the fault-free value.

Step 3: Compare the fault-free output values derived in Step 1 with those derived in Step 2. Provided there is at least one test input that produces different values, the fault is detected.

Step 4: Repeat Steps 2–3 for each of the other faults.

Solution to Part 3

Step 1: Create sensitive path from c_9 to c_{13}.
- a) Set $c_9 = 1$. c_9 is *fault generator* (s-a-0).
- b) Set $c_{12} = 0$ to enable fault-propagation through G5. $c_{12} = 0$ is a *fixed value*.
- c) Fault-effect now propagated to primary output z_1. c_{13} is a *fault-transmitter*.

Step 2: Consistency stage. Ensure $c_9 = 1$ and $c_{12} = 0$ values.
- a) $c_9 = 1$. This sets both $x_1 = 1$ and $x_2 = 1$.
- b) If $x_1 = 1$, then $c_{10} = 1$ and $c_{11} = 1$.
- c) If $c_{10} = 1$ and $c_{11} = 1$, then $c_{12} = 0$ as required.

Step 3: Conclusion and test.
Setting $x_1 = 1$ and $x_2 = 1$ establishes a fault-free value of 1 on c_9 and also the fixed value of 0 on c_{12}. If c_9 is s-a-0, the effect is propagated to z_1 along the sensitive path $c_9 \rightarrow c_{13}$.
The test therefore is $x_1 = 1, x_2 = 1$ with the expected fault-free value of $z_1 = 1$. The primary input x_3 is unassigned.

Exercise 2.1, and the outline solution, should help to consolidate understanding of the sensitive-path technique. It should also indicate something of the nature of a problem mentioned earlier—namely, that of deriving the individual and composite fault cover of the four tests. There are only two practical solutions to this problem. One is to use a logic simulator; the other is to insert faults physically onto the board, one by one, and run the test program. This topic is covered more fully in Chapter 4.

2.3.1 Unassigned Inputs

Before leaving this example, we will look at a further problem. The fifth test derived for the c_9 s-a-0 fault specified $x_1 = 1$ and $x_2 = 1$. The logic value on the third primary input, x_3, was not specified and, since either value is possible, it would seem that it does not matter what value is finally assigned. This is not strictly true, as Figures 2.6 and 2.7 demonstrate.

If x_3 is assigned 1 (Figure 2.6), then the resulting test covers four faults in the s-a-0 fault list. If x_3 is assigned 0 (Figure 2.7), the resulting

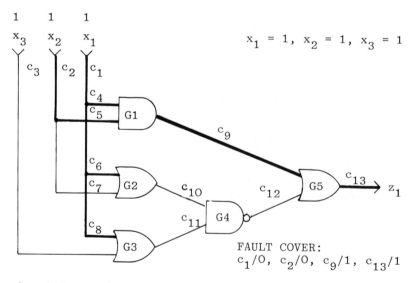

Sensitive paths shown bold

FIGURE 2.6 $x_3 = 1$

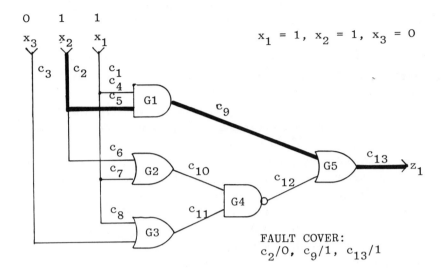

FIGURE 2.7 $x_3 = 0$

test covers only three faults. The reason for the difference is that the sensitive paths created by each test are slightly different, as shown in the figures. In Figure 2.6, a s-a-0 fault on c_1 is propagated through G1 and G5 and is therefore detectable. The fault also propagates to the input sides of both G2 and G3 via the fanout branches c_6 and c_8, but the values on primary inputs x_2 and x_3 block any further propagation through G2 and G3. This means that nodes c_{10} and c_{11} remain steady at 1, thereby keeping node c_{12} at 0, in order to maintain the sensitive path from c_9 to c_{13}.

Now observe what happens in Figure 2.7, where x_3 has been set to 0 instead of 1. A stuck-at-0 on c_1 will still propagate through G1 and change c_9 from 1 to 0, but it will also propagate through G3, changing c_{11} from 1 to 0. This change in c_{11} will now cause c_{12} to change from 0 to 1. The net effect on the primary output z_1 is that of no change. Either c_9 is 1 and c_{12} is 0 if there is no fault on c_1, or c_9 is 0 and c_{12} is 1 if there is a fault on c_1. In both cases therefore, one of the inputs to the OR gate, G5, is a 1, thereby holding the output at 1.

Fault-effects travelling along different sensitive paths are termed "multiple" sensitive paths. If the paths subsequently come together again at a common gate—G5 in this example—this is termed "fault-effect reconvergence." If the net outcome at the output of the reconverging gate is that of "no change" then this is called "negative" reconvergence. If the net outcome is that the output of the reconverging gate does change as a result of both inputs changing, then this is called "positive" reconvergence. In the example above, negative reconvergence occurs at gate G5 for the source fault c_1 s-a-0, and this is why this fault was not part of the fault cover for the test in which x_3 has been assigned 0. This question of reconvergence is an important one and we will return to it in the next section. Before we continue with the topic, however, we will address ourselves to one other question about Figures 2.6 and 2.7. The question is this: why does it matter what value we give to x_3? We only required a test for c_9 stuck-at-0 and, provided x_1 and x_2 are both equal to 1, we have satisfied this requirement.

The immediate answer is that it does not matter which of the two tests we choose from the point of view of detecting c_9 s-a-0. Other criteria may exist, however, on which we can base the assignment of values to unassigned variables. For example, if we were only interested in go/no go testing—that is, in fault detection only—then we may look for tests that have a high fault cover so as to keep the total number of tests down to a near-minimal level. If this were the case, the test with $x_3 = 1$ is better than the test with $x_3 = 0$.

Alternatively, if we were seeking to diagnose the board—that is, to detect and locate the fault—the $x_3 = 0$ test is better simply because the degree of uncertainty as to the cause of the fault is less than for the $x_3 = 1$ test. If the board fails the $x_3 = 1$ test, then the fault is one of four possible. If the board fails the $x_3 = 0$ test, the fault is one of three possible. Clearly the $x_3 = 0$ test is better from this point of view.

2.3.2 Fault-Effect Reconvergence

Returning to the subject of fault-effect reconvergence, consider Figure 2.8.

This diagram is a simpler demonstration of the two types of reconvergence—positive and negative. In the top half of the diagram, the only way a stuck-at-1 fault on node c_1 can be detected at the primary

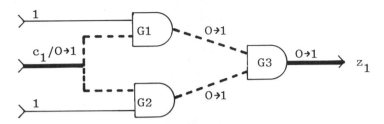

Positive reconvergence. The net outcome at
the reconverging gate, G3, is that the effect
of the fault continues to propagate on
through the gate.

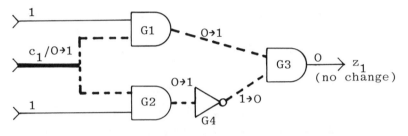

Negative reconvergence. The net outcome at
G3 is that the output does not change. The
propagation of the fault-effects is blocked
at G3.

FIGURE 2.8 Positive and Negative Reconvergence

output z_1 is to propagate the fault through both G1 and G2 and from there
to the reconverging gate, G3. The polarities of the logic value changes
on the two inputs to G3 are such that the output of G3 also changes,
thereby making the original fault observable at z_1. This is positive
reconvergence.

The bottom part of Figure 2.8 shows the converse situation. The
presence of the inverter at G4 causes the two reconverging inputs to G3
to have opposite polarities, and the net effect at z_1 is no change. (Is it
possible to detect a stuck-at-1 fault on c_1 at the output z_1?)

Figures 2.9 and 2.10 illustrate a more realistic example of the
reconvergence phenomena.

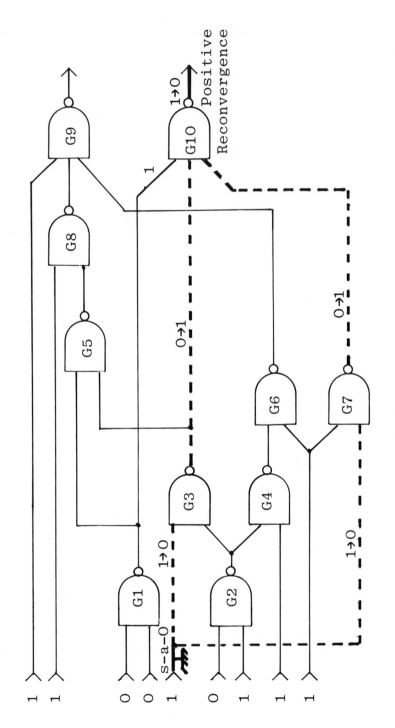

FIGURE 2.9 Positive Reconvergence

34

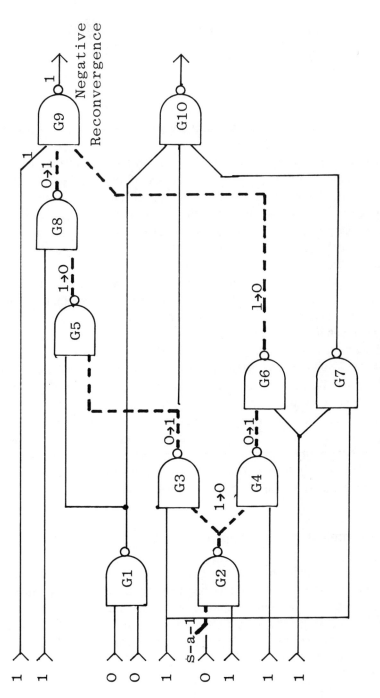

FIGURE 2.10 Negative Reconvergence

35

In Figure 2.9, the s-a-0 fault shown on the trunk of the fanout point is detected at the lower output by virtue of the positive reconvergence at gate G10. Note that the reconvergence is on a subset of the inputs to this gate. The other input is fixed at 1 (its enable level).

In Figure 2.10, negative reconvergence occurs at gate G9 for the source s-a-1 fault shown.

The observant reader will notice that, in the case of the positive reconvergence, the even or odd polarities of the number of signal inversions along each path from the source of the fault to the reconverging gate are the same (both odd in this case). For the negative reconvergence example, the polarities are different (one even, one odd). This observation can form the basis for an algorithm to determine the nature of the fault-effect reconvergence.

The reason an understanding of reconvergence is important is that, in circuits possessing feedback from one device to another, reconvergence can take place "around the loop." When this happens, reconvergence can destroy various assumptions concerning fixed values that have been established either to help set up the fault-generator value or to create the sensitive path. Both manual and algorithmic analysis of this problem can be complex and time-consuming, particularly if the loop contains stored-state devices such as flip-flops or monostables. For these devices, the flow of fault information is time dependent as well as device dependent, and this may mean that, for certain faults in the fault list, it is not possible in the time allowed and with the resources available to generate a suitable test. This can mean that the final fault cover falls somewhat short of the target figure of 100% of all faults in the fault list.

2.4 Example 4: Fault-collapsing

We turn now to a side issue in test-pattern generation—that of reducing the number of faults in the fault list before any generation of test-patterns actually takes place. This reduction is carried out by analyzing the circuit structure (gate types and their interconnection) while looking for certain test-fault relationships. The process is called "fault collapsing."

Consider the 3-input AND gate shown in Figure 2.11.

Intuitively, we see that the only way to test for the output connection c_4 s-a-0 is to set all the inputs high, i.e., $x_1 = x_2 = x_3 = 1$, thus

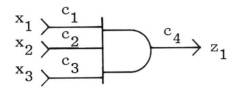

FIGURE 2.11 And Gate

causing the output to go high if the circuit is fault-free. Any other input combination would set the output low and not be a test condition for this fault. A further study of this particular test, however, reveals that it is also the only way to test for c_1 s-a-0, c_2 s-a-0, or c_3 s-a-0. This means that if the test fails, i.e., if the output value monitored is 0, then there is no further test that can be applied to distinguish which of the four faults has occurred (short of actual probing on the connections themselves). In other words, if a test is provided for any one of the three input s-a-0 faults, it will also test for the output s-a-0 and vice versa. These four faults are totally indistinguishable, and the relationships are written as shown below, using the shorthand notation $c_1/0$ to mean c_1 s-a-0, etc.

$c_1/0 <=> c_4/0$

$c_2/0 <=> c_4/0$ producing $c_1/0 <=> c_2/0 <=> c_3/0 <=> c_4/0$

$c_3/0 <=> c_4/0$

    The double-headed arrow ($<=>$) indicates the bidirectional nature of the implication. The relationship is called "2-way."

    Consider now a test for the output c_4 s-a-1. There are in fact seven tests in all, i.e., the seven combinations that set c_4 to 0, but note that if an input combination is selected that sets one input low and the other two high, the test will also cover the s-a-1 fault on the low input, e.g., $x_1 = 0$, $x_2 = x_3 = 1$ covers c_1 and c_4 s-a-1. If the circuit fails this test, it is possible to distinguish further between the two faults by applying another test: $x_1 = 1$, $x_2 = 0$, $x_3 = 1$, for instance. In other words, although a test for c_1 s-a-1 must also test for c_4 s-a-1, the reverse is not true. There are three such test-fault implication relationships in the circuit. They are written:

$$c_1/1 \Longrightarrow c_4/1$$

$$c_2/1 \Longrightarrow c_4/1$$

$$c_3/1 \Longrightarrow c_4/1$$

The single-headed arrow ($\Longrightarrow$) indicates the unidirectional nature of the test-fault implication. The relationship is called "1-way."

Similar results apply to other basic gates but, before demonstrating the usefulness of fault-collapsing, we will consider the test-fault implication of fanout connections.

Fanout can be classified as reconvergent or non-reconvergent dependent upon whether some or all the fanout branches subsequently reconverge at a later gate or do not converge. For non-reconvergent fanout (which must imply that the circuit is multi-output), a test for a s-a-1 or s-a-0 fault on a particular branch must also test for the same fault on the trunk, but not necessarily vice versa. This is because the effect of the fault on the trunk may travel down some branch other than the branch in question. This is illustrated in Figure 2.12.

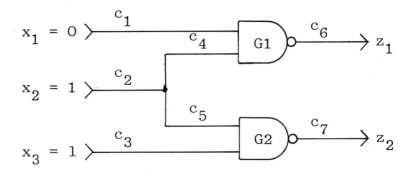

FIGURE 2.12 Non-Reconvergent Fanout

For this circuit, a test for $c_4/0$ or $c_4/1$ is also a test for $c_2/0$ or $c_2/1$ respectively, i.e., $c_4/0 => c_2/0$ and $c_4/1 => c_2/1$. The relationship is not bidirectional, however: a test for $c_2/0$ does not necessarily imply a test for $c_4/0$. For example $x_1 = 0$, $x_2 = x_3 = 1$ will test for $c_2/0$ but not for $c_4/0$. At best therefore, the general implication relationship for non-reconvergent fanout is that it is 1-way from the branch to the trunk.

For circuits with reconvergent fanout the situation is more complex, as illustrated by the negative reconvergent situation shown again in Figure 2.13.

This circuit presents a simple counterexample to the premise that a test for a fanout branch s-a-1 or s-a-0 automatically provides a test for the same fault on the trunk irrespective of the presence or absence of reconvergent fanout. For the input condition shown, the effect of a s-a-1 fault on the branch c_4 is propagated through G1 and G4 to change the output value on c_9 (assuming c_5 retains its correct value of 0). This input condition is therefore a test for $c_4/1$, but it is not a test for the same fault on the trunk, i.e., for $c_2/1$. This is because of the negative reconvergence effect as described in the previous section, i.e., $c_2/1$ travels down both branches in parallel, changing c_6, c_7, and c_8 in such a way that c_6 and c_8 "cancel" each other out at gate G4. The net effect is that c_9 remains unchanged and that $c_2/1$ is not detected by the input condition.

Unfortunately, as in the case of non-reconvergent fanout, the converse is also not generally true. A test for the trunk s-a-1 or s-a-0 is not necessarily a test for any single branch s-a-1 or s-a-0. In fact, if the inverter G3 is removed from Figure 2.13, and c_7 and c_8 connected together (the positive reconvergence situation), it can be seen that the same test input now becomes a test for the trunk fault $c_2/1$ but is not a test

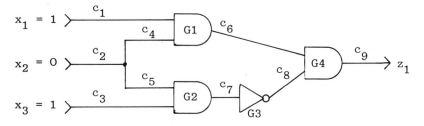

FIGURE 2.13 Reconvergent Fanout

for either of the single branch faults $c_4/1$ or $c_5/1$, assuming that these faults do not affect the other branches or the trunk.

From these two examples, therefore, we conclude that there is no simple rule for 1-way or 2-way test-fault relationships for fanout nodes that subsequently reconverge.

Figure 2.14 summarizes the relationships for basic gates and for non-reconvergent fan-out points.

GATE TYPE	2-WAY IMPLICATIONS	1-WAY IMPLICATIONS
AND	Any input/0 ⟷ Output/0	Any input/1 ⟹ Output/1
OR	Any input/1 ⟷ Output/1	Any input/0 ⟹ Output/0
NAND	Any input/0 ⟷ Output/1	Any input/1 ⟹ Output/0
NOR	Any input/1 ⟷ Output/0	Any input/0 ⟹ Output/1
INV	Input/0 ⟷ Output/1	--
	Input/1 ⟷ Output/0	
NONRECONVERGENT FAN-OUT POINT	—	Any branch/0 ⟹ Trunk/0
		Any branch/1 ⟹ Trunk/1

FIGURE 2.14 2-Way and 1-Way Implications

To illustrate how the concepts of test-fault relationships can be used to reduce the number of faults in a fault list, consider the circuit shown in Figure 2.15 (Example 4).

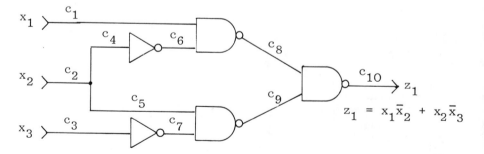

$$z_1 = x_1\bar{x}_2 + x_2\bar{x}_3$$

FIGURE 2.15 Example 4

This circuit contains 10 individual connections, c_1 to c_{10}, and therefore 20 single stuck-at fault conditions. The 2-way and 1-way relationships are as follows:

2-way relationships
$c_4/1 <=> c_6/0$
$c_4/0 <=> c_6/1$
$c_3/1 <=> c_7/0$
$c_3/0 <=> c_7/1$
$c_1/0 <=> c_6/0 <=> c_8/1$
$c_5/0 <=> c_7/0 <=> c_9/1$
$c_8/0 <=> c_9/0 <=> c_{10}/1$

1-way relationships
$c_1/1 => c_8/0$
$c_6/1 => c_8/0$
$c_5/1 => c_9/0$
$c_7/1 => c_9/0$
$c_8/1 => c_{10}/0$
$c_9/1 => c_{10}/0$

Considering just the 2-way implications, we see that in one set of faults, $c_6/0$ is equivalent to both $c_1/0$ and $c_8/1$ whereas, in another set, $c_6/0$ is equivalent to $c_4/1$. Therefore the two sets of faults containing $c_6/0$ can be combined into a single, larger set of equivalent faults, given by $\{c_1/0, c_4/1, c_6/0, c_8/1\}$. Examining all the 2-way sets in an exhaustive fashion reduces the seven starting sets to the following five sets:

S1 : $\{c_3/0, c_7/1\}$
S2 : $\{c_4/0, c_6/1\}$
S3 : $\{c_1/0, c_4/1, c_6/0, c_8/1\}$
S4 : $\{c_3/1, c_5/0, c_7/0, c_9/1\}$
S5 : $\{c_8/0, c_9/0, c_{10}/1\}$

These five sets cover all but five of the single stuck-at fault conditions. The five not covered are listed in the following five single-member sets:

S6 : $\{c_1/1\}$
S7 : $\{c_2/1\}$
S8 : $\{c_2/0\}$
S9 : $\{c_5/1\}$
S10 : $\{c_{10}/0\}$

On the basis of the 2-way relationships above, we are able to reduce the number of faults in the fault list from 20 to 10, a reduction of 50%. The reduced fault list now contains one fault, selected arbitrarily, from the 10 sets of equivalent faults listed above, S1 to S10.

A further reduction can be achieved, using the 1-way relationships, by observing that:

(i) a test for detecting $c_1/1$ (S6) will also detect $c_8/0$ and also, any faults indistinguishable from $c_8/0$, i.e., all those faults listed in S5.

(ii) a test for detecting $c_8/1$ (S3) or $c_9/1$ (S4) will also detect $c_{10}/0$, i.e., set S10.

Fault sets S5 and S10 can therefore be removed from the list of fault-sets to be considered, thereby reducing the number to 8.

This represents a final reduction in the size of the fault list of 60% and demonstrates the power of using test-fault relationships either prior to test-pattern generation or test-pattern evaluation.

The following exercise contains a further example of fault collapsing.

Exercise 2.2

Carry out a fault-collapsing exercise on the circuit shown in Figure 2.16.

Exercise 2.2: Outline Solution

2-way relationships

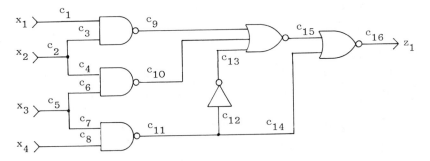

FIGURE 2.16 Exercise 2.2

$c_1/0 <=> c_3/0 <=> c_9/1$ $c_1/0, c_3/0, c_9/1, c_4/0, c_6/0,$

$c_4/0 <=> c_6/0 <=> c_{10}/1$ $c_{10}/1, c_{12}/0, c_{13}/1, c_{15}/0$

$c_7/0 <=> c_8/0 <=> c_{11}/1$

$c_{12}/0 <=> c_{13}/1$

$c_{12}/1 <=> c_{13}/0$

$c_9/1 <=> c_{10}/1 <=> c_{13}/1 <=> c_{15}/0$

$c_{15}/1 <=> c_{14}/1 <=> c_{16}/0$

1-way relationships

$c_1/1 => c_9/0$

$c_3/1 => c_9/0$

$c_4/1 => c_{10}/0$

$c_6/1 => c_{10}/0$

$c_7/1 => c_{11}/0$

$c_8/1 => c_{11}/0$

$c_9/0 => c_{15}/1$

$c_{10}/0 => c_{15}/1$

$c_{13}/0 => c_{15}/1$

$c_{15}/0 => c_{16}/1$

$c_{14}/0 => c_{16}/1$

Consideration only of the 2-way relationships produces the following
set of collapsed equivalent faults and remaining single-member sets.

$S1 = \{c_1/0, c_3/0, c_9/1, c_4/0, c_6/0, c_{10}/1, c_{12}/0, c_{13}/1, c_{15}/0\}$
$S2 = \{c_7/0, c_8/0, c_{11}/1\}$
$S3 = \{c_{12}/1, c_{13}/0\}$
$S4 = \{c_{15}/1, c_{14}/1, c_{16}/0\}$
$S5 = \{c_1/1\}$
$S6 = \{c_2/0\}$
$S7 = \{c_2/1\}$
$S8 = \{c_3/1\}$
$S9 = \{c_4/1\}$
$S10 = \{c_5/0\}$
$S11 = \{c_5/1\}$
$S12 = \{c_6/1\}$
$S13 = \{c_7/1\}$
$S14 = \{c_8/1\}$
$S15 = \{c_9/0\}$
$S16 = \{c_{10}/0\}$
$S17 = \{c_{11}/0\}$
$S18 = \{c_{14}/0\}$
$S19 = \{c_{16}/1\}$

Further consideration of the 1-way relationships eliminates the following sets:

(i) S15. $c_9/0$ is covered by $c_1/1$ (S5) or $c_3/1$ (S8)
(ii) S16. $c_{10}/0$ is covered by $c_4/1$ (S9) or $c_6/1$ (S12)
(iii) S17. $c_{11}/0$ is covered by $c_7/1$ (S13) or $c_8/1$ (S14)
(iv) S4. $c_{15}/1$, and any fault equivalent to it, is covered by $c_9/0$ (S15), $c_{10}/0$ (S16), or $c_{13}/0$ (S3). The first two faults have already been eliminated ((i) and (ii) above), so the test for $c_{15}/1$ is now guaranteed by a test for S5, S8, S9, and S12, or by S3 (covering $c_{13}/0$).
(v) S19. $c_{16}/1$ is covered by $c_{15}/0$ (S1) or $c_{14}/0$ (S18).

The final fault list consists of 14 stuck-at-faults selected from the 14 remaining sets. This represents a reduction of 56% on the original list of 32 faults. .

Chapter 3
Test-Pattern Generation:
Sequential Circuits

This chapter continues the intuitive introduction to test-pattern generation but extends it into circuits containing stored state devices and, eventually, global feedback. Before continuing with the main series of examples, we pause to reflect further upon the true nature of sensitizing paths through a device. .

3.1 The Nature of Sensitive Paths

The process of sensitizing a path from the input side of a device to one or more of the outputs is central to any fault-oriented procedure for generating test patterns. In general, a sensitive path exists from a device input, x, to a device output, z, if, and only if, the effect of a change in the logical value of x can be observed at z. The change in question can be due to the node x being either a fault generator itself or a fault transmitter of an earlier fault. Notice however, that the observation at z does not necessarily imply that z has to change as a result of the change in x. The observation, from z, of the change in x can be deduced from the fact that an expected change in z did not take place. The following series of examples (Figures 3.1 to 3.4) illustrate the many forms of path sensitization through logic devices.

Figure 3.1 (a) shows the simplest form of a sensitive path. The value on the output connection, c_4, is controlled by the 0 on c_1, provided that c_2 and c_3 maintain their fault-free values of 1. A generated stuck-at-1 fault on c_1, or a transmitted 0 to 1 change (written $c_1/0{\rightarrow}1$) will be propagated to the output c_4 ($c_4/0{\rightarrow}1$) and thence on through any following successor devices.* The only single input-to-output sensitive path that exists through the gate is from c_1 to c_4. This path may be written as the ordered pair (c_1,c_4).

Figure 3.1(b) shows a similar condition on a 3-input OR gate. In this case, a change of logic level on any of the input connections will be reflected at the output connection, and we can identify three separate sensitive paths: (c_1,c_4), (c_2,c_4), and (c_3,c_4).

Figure 3.1(c) shows a multi-function element—a 3-input OR-NOR gate. The logic levels on the input connections are such as to establish c_2 as a control input for both the output connections c_4 and c_5. $c_2/1{\rightarrow}0$ causes $c_4/1{\rightarrow}0$ and $c_5/0{\rightarrow}1$ simultaneously and the sensitive paths are (c_2,c_4) and (c_2,c_5).

Figure 3.1(d) shows an exclusive-OR (EOR) gate. Both inputs are control inputs; the sensitive paths are (c_1, c_3) and (c_2, c_3). This example

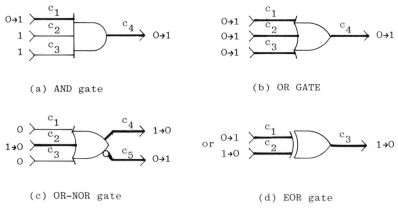

(a) AND gate

(b) OR GATE

(c) OR–NOR gate

(d) EOR gate

FIGURE 3.1 Sensitive Paths

*The notation $c_1/0{\rightarrow}1$ and $c_4/0{\rightarrow}1$ should be read as "c_1 changing from 0 to 1" causes "c_4 to change from 0 to 1." In other words, the interpretation of the slash (/) varies according to the cause or effect status of the change of value.

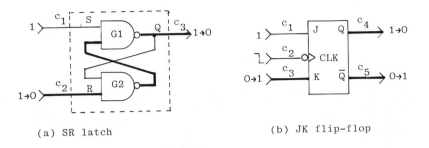

(a) SR latch　　　　　　　　　　　　(b) JK flip-flop

FIGURE 3.2 Stored-State Devices

illustrates an important property of the EOR gate, namely, that sensitive paths always exist from both input connections to the output connection for all combinations of input logic levels.

Figure 3.2(a) shows an SR latch assumed to be constructed from two cross-coupled NAND gates (shown inside the dotted boundary). If we consider the element at the latch level, rather than at the equivalent-circuit gate level, then, for the logic levels shown and an initialized value of $Q = 1$, a sensitive path exists from the R input connection c_2 to the Q output connection c_3—this being the only sensitive path.

If we consider the element at its equivalent-circuit level then we note that there are two gate-level sensitive paths, one through G1 and the other through G2. These two gate-level paths can be chained together to produce the final element-level path (c_2, c_3), but we must take care that the feedback path from G1 output to G2 input does not invalidate this element-level path. (This, one of the major problems in applying the sensitive path-technique to sequential circuits, is further demonstrated later in the chapter.)

Figure 3.2(b) depicts a JK flip-flop whose output levels have been initialized to $Q = 1$, $\overline{Q} = 0$. A change in the logic level on the J input $(c_1/1 \rightarrow 0)$ will not change either of the output values, but a change in the logic level on the K input $(c_3/0 \rightarrow 1)$ will do so, provided a suitable clock transition occurs. In other words, c_3 is a control input establishing the conditions for a sensitive path through the element, but the actual creation of the path itself is dependent on, and controlled by, a change in the clock input. Such a path is sometimes called a "dynamic" path and can be differentiated from the "static" paths of the previous examples. A static path has the property that the path is always there and is activated immediately a change of input level occurs. The dynamic path, on the other hand, only exists for a short time, i.e., during the clock transition.

The JK example can also be used to illustrate the observation, made in the sensitive-path definition, regarding the absence of an expected change in the output. Consider the situation where $c_1 = 1$ as before, but where $c_3 = 1$ rather than 0. This is the toggle mode and, under normal conditions, the next clock pulse would cause the flip-flop to change state, i.e., $c_4/1{\rightarrow}0$ and $c_5/0{\rightarrow}1$. If $c_3/1{\rightarrow}0$ before the clock, then the flip-flop will not change state and the effect of $c_3/1{\rightarrow}0$ is observed as $c_4/1{\rightarrow}1$ and $c_5/0{\rightarrow}0$, i.e., by the absence of the expected change.

The fact that sensitive paths through clock-controlled logic elements are dynamic is another major problem in generating test patterns for sequential circuits. This aspect of test generation is also demonstrated later in the chapter.

Figure 3.3 shows a more complex combinational circuit element—a 4-way multiplexer. With just a knowledge of its functional

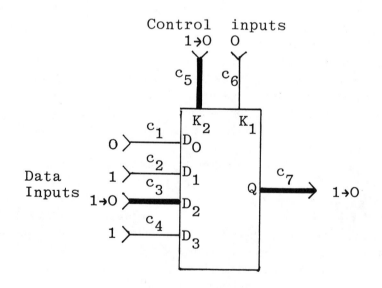

FIGURE 3.3 Multiplexer

behavior, we can see that with the selection code shown ($K_2 = 1$, $K_1 = 0$) the output Q is controlled by the input D_2, i.e., (c_3,c_7) is a sensitive path. Note also that another path exists from one of the control inputs K_2 to Q. If $K_2/1 \rightarrow 0$, the data input selected will change from D_2 to D_0 causing the output $Q/1 \rightarrow 0$. (c_5,c_7) is also a sensitive path, and, in both cases, the paths are static.

Figure 3.4 shows a more complex synchronous sequential circuit—a 3-stage serial-in, parallel-out shift register constructed from D-type flip-flops and assumed to be initialized to logic 1 in all stages. As in the previous synchronous sequential circuit (the JK flip-flop), the sensitive paths are dynamic, and the effect of the change $c_1/1 \rightarrow 0$ on the serial path input line is observable at all three of the parallel data output lines, although not directly so. The effect of $c_1/1 \rightarrow 0$ propagates through each stage of the shift register as successive clock pulses are applied. This timed delay of fault-effect propagation introduces a further complication into test-pattern generation for synchronous sequential circuits.

Finally, before leaving these examples, we make two further observations about the nature of sensitive paths.

The first observation concerns the single or dual-purpose nature of the path. Consider again Figure 3.1(a). It was established that (c_1, c_4) was a sensitive path for which $c_1/0 \rightarrow 1$ caused $c_4/0 \rightarrow 1$. If c_1 now changes from 1 to 0 again, then c_4 will change correspondingly. The sensitive path is still valid but this time it is valid for the opposite changes in the polarity of the logic levels on c_1 and c_4, i.e., $c_1/1 \rightarrow 0$ causes $c_4/1 \rightarrow 0$. This type of path, true for both initial polarities on the input terminal, is called a ''dual-polarity'' sensitive path. (The situation

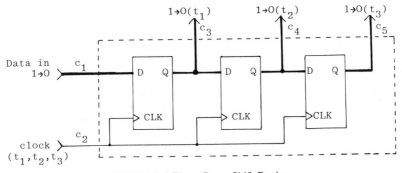

FIGURE 3.4 Three-Stage Shift Register

is analogous to the controlled movement of an arm, say, on a puppet. If the puppeteer pulls the control string up, the arm goes up. If the string is now relaxed, the arm falls back. This cycle can be repeated indefinitely.)

Not all paths in Figures 3.1 to 3.4 have this property however. Consider the S-R latch (Fig. 3.2(a)). If $c_2/1 \rightarrow 0$, then certainly $c_3/1 \rightarrow 0$, assuming the initialized value of 1 as before. Unfortunately, the new value of 0 on c_3 will latch G2 output to 1, thereby holding c_3 at 0 irrespective of any further change on c_2. The path is therefore valid only for a single polarity change in the value of c_2, rather than for both changes of polarity. Such a path is termed a "single-polarity" path. (Continuing the analogy of the puppet's arm, it is as though having raised the arm by pulling the string, the arm now locks into position and cannot be lowered by releasing the string. Other control forces have to be employed to lower the arm.)

The second observation concerns the local or global nature of a path. In the consideration of the SR latch (Figure 3.2(a)), it was suggested that the (c_2, c_3) sensitive path could be derived by chaining together two gate-level sensitive paths—one through G_1, the other through G_2. This chaining process is fundamental to the sensitive-path technique and it is sometimes convenient to refer to sensitive paths in terms of their local or global status depending on how the element itself is viewed. If the element is considered a basic unit in its own right and is not to be analyzed in terms of an equivalent circuit, then any sensitive path through the element is called a "local" path. If the element is considered in terms of an equivalent circuit using "lower order" elements and the derivation of the element sensitive path is based on an analysis of the paths in the equivalent circuit, then the path is called a "global" path.

As far as the SR latch is concerned, (c_2, c_3) is a local path if the latch is considered as a single unit whose logical function is described by a characteristic equation. If the latch is considered to be a circuit consisting of two cross-coupled NAND gates however, then a local sensitive path exists through each NAND gate and these paths are concatenated to produce the global sensitive path (c_2, c_3).

Obviously, sensitive paths through any of the primitive logic gates AND, OR, NAND, NOR, or INV must always be local paths.

For the sake of brevity, the status of a sensitive path (static or dynamic, single-polarity or dual-polarity, local or global) will not be defined unless necessary to clarify a point.

Exercise 3.1 contains further examples of sensitive paths through a stored-state device.

Exercise 3.1

Figure 3.5 shows an SN 7473 J-K flip-flop and transition table. For each of the four inputs, J,T,K, and C, determine the full set of sensitive paths that will allow observation of stuck-at faults at the Q output. Use the convention T = 1 to signify the presence of a negative clock edge and T = 0 to signify the absence of the negative clock edge. In this respect therefore, the faults on the clock are considered to be ''no negative edge generated'' (T/1 → 0) and ''negative edge generated'' (T/0 → 1), rather than the more usual s-a-0, s-a-1.

Exercise 3.1: Outline Solution

Consider Figure 3.6. Fault effect propagation takes place if the presence of the fault causes the output to change where it should not have done so, or if the output does not change where it should have done so. Both circumstances allow detection of the fault condition. For this reason, the expected output change has been included in Figure 3.6.

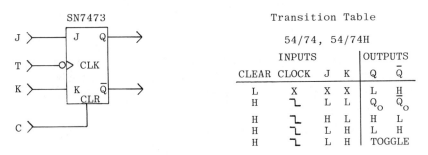

SN7473

Transition Table

54/74, 54/74H

INPUTS				OUTPUTS	
CLEAR	CLOCK	J	K	Q	$\bar{Q}$
L	X	X	X	L	H
H	⌐	L	L	Q_O	$\bar{Q}_O$
H	⌐	H	L	H	L
H	⌐	L	H	L	H
H	⌐	L	H	TOGGLE	

FIGURE 3.5 SN7473 J-K Flip-Flop

| Initial state | | | | | Fault-free | Fault detected at Q output? | | | |
J	K	T	C	Q	response	J input	K input	T input	C input
0	0	0	0	0	NC	N	N	N	N
0	0	0	0	1	*	*	*	*	*
0	0	0	1	0	NC	N	N	N	N
0	0	0	1	1	NC	N	N	N	N
0	0	1	0	0	NC	N	N	N	N
0	0	1	0	1	*	*	*	*	*
0	0	1	1	0	NC	Y	N	N	N
0	0	1	1	1	NC	N	Y	N	Y
0	1	0	0	0	NC	N	N	N	N
0	1	0	0	1	*	*	*	*	*
0	1	0	1	0	NC	N	N	N	N
0	1	0	1	1	NC	N	N	Y	Y
0	1	1	0	0	NC	N	N	N	N
0	1	1	0	1	*	*	*	*	*
0	1	1	1	0	NC	Y	N	N	N
0	1	1	1	1	C	N	Y	Y	N
1	0	0	0	0	NC	N	N	N	N
1	0	0	0	1	*	*	*	*	*
1	0	0	1	0	NC	N	N	Y	N
1	0	0	1	1	NC	N	N	N	Y
1	0	1	0	0	NC	N	N	N	Y
1	0	1	0	1	*	*	*	*	*
1	0	1	1	0	C	Y	N	Y	Y
1	0	1	1	1	NC	N	Y	N	Y
1	1	0	0	0	NC	N	N	N	N
1	1	0	0	1	*	*	*	*	*
1	1	0	1	0	NC	N	N	Y	N
1	1	0	1	1	NC	N	N	Y	Y
1	1	1	0	0	NC	N	N	N	Y
1	1	1	0	1	*	*	*	*	*
1	1	1	1	0	C	Y	N	Y	Y
1	1	1	1	1	C	N	Y	Y	N

Legend: NC = No Change
          C = Change
          Y = Yes
          N = No
          * = Invalid start state

FIGURE 3.6 Sensitive Paths

Each entry can now be evaluated, against these criteria, by the simple expedient of altering the polarity of the defined value on each bona-fide input and observing the resulting effect at the Q (and $\overline{Q}$) output.

For example, the J = 0, K = 1, T = 1, C = 1, Q = 1 state is evaluated for J s-a-1, K s-a-0, T no edge generated, C s-a-0 respectively. The fault-free outcome of this state is that the Q output will change (from 1 to 0). If J is s-a-1, then the Q output will still change (the flip-flop is now in its toggle mode) and the fault is not detected because the faulty response is identical to the fault-free response.

If the K input is s-a-0, however, the Q output will not change and the fault is detected. Similarly for T no edge generated. Finally if C is s-a-0 then this will cause Q to change. Again, this is what was expected, so the fault is not detected.

The other point to note about this evaluation is that eight of the 32 possible start states are actually invalid. These eight states correspond to the C = 0, Q = 1 state. A value of C = 0 will force $\overline{Q}$ to 1 and hence Q to 0. The absence of a preset input means that Q will always go to 0 if C is 0. Hence any combination of C = 0, Q = 1 is invalid.

The main implication of this is that the total number of states is not 32 but (32 − 8) = 24. In other words, the total number of ways of either sensitizing or not a particular input to the Q output is equal to the total number of valid states, and not to the apparent total number of states.

3.2 Example 5 : Stored-State Devices

Example 5, shown in Figure 3.7, is the circuit previously discussed in Chapter 1.

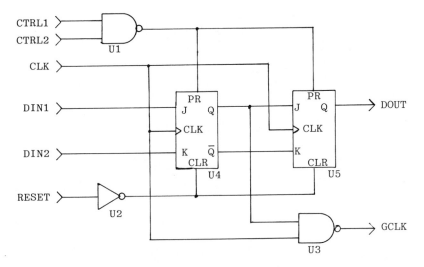

FIGURE 3.7 Example 5

The test strategy for the circuit, repeated again in the timing diagram of Figure 3.8, is based wholly on the functional approach.

What is not yet known is how effective the sequence of input pin changes will be in creating test conditions for stuck-at faults on the nodes. What we suspect is that a functional checkout of the sort defined by the timing diagram will catch most, if not all, of the stuck-at faults. Without a full evaluation however, we are unable to verify this conjecture. If the sequence proves capable of detecting all the faults in the fault list (all nodes stuck-at-1 or stuck-at-0, say), then we have met the first objective of testing. If this is not the case, then we will probably have succeeded in detecting most of the faults anyway and need only concentrate on the few remaining. We will return to this line of thought later in this chapter.

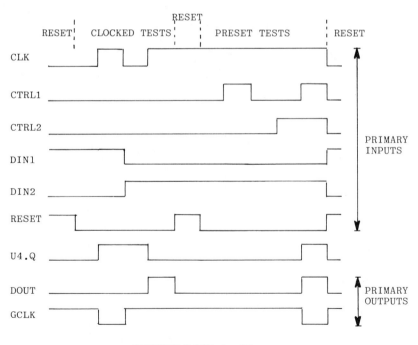

FIGURE 3.8 Timing Diagram

3.3 Example 6 : Initialization

The next example, Example 6, illustrates yet another possible problem. The circuit, shown in Figure 3.9, contains both stored-state devices and global feedback and is, in fact, a self-starting divide-by-5 counter.

The count sequence, defined by the state diagram, is shown in Figure 3.10. The entries within each state represent the values on the three outputs A, B, and C. The main count sequence is shown by the inner group of five states, with transitions between states controlled by the negative edge of the CLOCK input.

What is missing from this circuit is the facility to clear the state of the counter to the all-0's state before commencing the count, that is, there is no general CLEAR line. From a logic design point of view, this is of no consequence because the circuit was designed to be self-starting. If, on power-up, the state of the circuit is not one of the main count-sequence states, then the arrival of the first negative-edge on the CLOCK will take the circuit into one of the five valid count states as shown. The normal sequence of states will then follow.

From a testing point of view, this is not desirable simply because the circuit cannot be placed into a known start state prior to testing. In other words, initializing the circuit, although not difficult, is more complex than it need be. One solution is to power-up the circuit, start

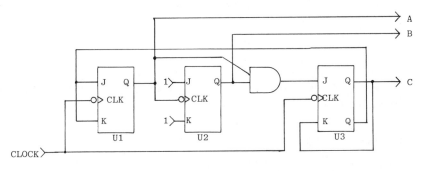

Self-starting divide-by-5 counter.

FIGURE 3.9 Example 6

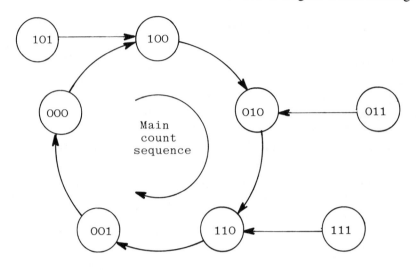

Note:

(i)   Values inside each state (circle) =
        values for A, B & C respectively.

(ii)   Transition between states occurs
        on the negative-edge of CLOCK.

(iii)   States 101, 011 and 111 are invalid.

FIGURE 3.10 State Diagram

clocking the counter, and program the tester to recognize a particular set of values on A, B, and C (A = B = C = 0, say). Modern test-programming languages contain conditional-type statements that allow the tester to test the value of an on-board signal, or set of signals, and branch accordingly. Alternatively, of course, we can request that the logic designer provide a master CLEAR line to initialize the three flip-flops to 0. This approach removes the problem rather than solves it.

Example 6 has illustrated one of the ways in which the logic designer can create problems for the test programmer. As will be discussed in Chapter 9, there are many more such ways.

3.4 Example 7 : Importance of Layout

This example, shown in Figure 3.11, illustrates a further problem in writing test patterns.

The diagram shows an asynchronous sequential circuit with five inputs and two outputs. The problem is—where do you start in writing a test program? The circuit appears to have no recognizable logic function. Also, there appear to be just as many lines going from right-to-left as from left-to-right. Because the overall function is obscure, the only viable strategy to generate the test patterns appears to be based totally on the structural fault-oriented approach, i.e., set up a fault list of all nodes stuck-at-1 and stuck-at-0 and working from the top of the list, use the sensitive-path technique to generate tests.

Figure 3.12 shows the same circuit as Figure 3.11 but drawn in a different way.

This time, its function is recognizable. It is, in fact, an equivalent NAND gate circuit for a master slave J-K flip-flop. The test strategy now becomes obvious: test it out as a J-K flip-flop in the usual way, i.e.,

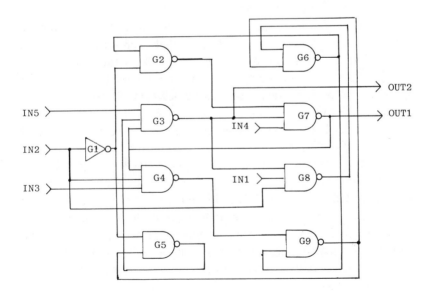

FIGURE 3.11 Example 7

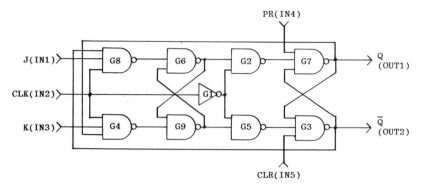

FIGURE 3.12 Example 7 (Redrawn)

test preset facility; test clear facility; and test clocked behavior using four sets of data changes on the J and K inputs. This example illustrates that visual layout of the circuit for which tests are to be generated can be very important. If the circuit diagram is presented on a disjointed series of sheets of paper, rather than as a single schematic, then it is necessary to re-create the complete circuit diagram. To carry out this process, the following set of rules will help.

RULE 1. Maximize flow of logic signal values from left to right. That is, minimize the number of lines that represent reverse flow (right-to-left) paths, such as the output of G6 to the input of G2 in Figure 3.11. In Figure 3.12 we see that this path is not a feedback path. (The following section explains the difference between "reverse flow" and "feedback.")

RULE 2. Place devices close together if there is considerable information flow between them. The term "considerable" is relative here, but as a simple example of the rule, G6 and G9 in Figure 3.11 are not that close together, yet the output of G6 goes to G9 and the output of G9 goes to G6. In Figure 3.12, G6 and G9 are closer and the joint function they perform, that of a latch, is easier to recognize.

RULE 3. Place devices whose inputs are primary inputs to the left, and devices whose outputs are primary outputs to the right. This rule is really a derivative of the first rule (maximizing left-to-right signal flow), and is illustrated by the positions of G8 and G3 in Figure 3.11. In Figure 3.12, G8 has been placed farther to the left and G3 farther to the right. In general this rule is not as powerful as the first two rules largely because primary inputs and primary outputs tend to enter and leave a logic circuit at many points. The rule is useful, however, as a means of arbitrating between choices that satisfy Rules 1 and 2.

RULE 4. Keep the number of cross-overs between signal lines to a minimum. This tends to ease the visual process of following a line from one device to another. If cross-overs cannot be avoided, then the use of different colored lines is recommended.

3.4.1 Reverse Flow and Feedback

Figure 3.13 illustrates the difference between the terms ''reverse flow'' and ''feedback.''

Figure 3.13(a) shows the traditional way of drawing a latch comprised of two cross-coupled NAND gates. In this diagram, the two output-back-to-input connections are drawn right-to-left. In this sense, they indicate a reverse flow of logic information. In reality, however, there is only one feedback path in the circuit, as shown by the redrawn version in Figure 3.13(b)*. From the point of view of recreating the circuit for test-pattern generation, Figure 3.13(b) is preferable to Figure 3.13(a) (although it is easier to identify the function in Figure 3.13(a)). The main objective of the layout rule, Rule 1, is to reduce the number of reverse-flow lines to the minimum value equal to the number of real

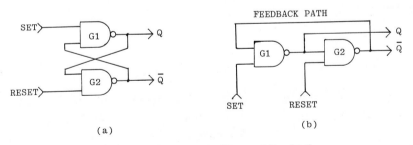

(a)   (b)

FIGURE 3.13 Reverse-Flow and Feedback

*The author has often put Figure 3.13(a) on a screen and asked ''How many feedback paths?'' On many occasions the answer has been ''two,'' even from an audience of digital engineers, not just from students!

feedback paths in the circuit. In this way, the test programmer is not confused by the presence of apparent feedback.

3.5 Example 8 : Problems of Choice

The next example, Example 8, is shown in Figure 3.14. Example 8 is considerably more complex than any of the previous examples and is representative of a small, but actual printed-circuit board.

Figure 3.15 shows the truth tables for the SN7442 and SN7473 devices used on this board.

The reason for including this example is to illustrate a problem that occurs quite frequently in generating test patterns for specific faults. This problem is that of choice. Very often, in generating a test, there is a choice of possibilities: a choice of paths along which the fault effect can be propagated; a choice of fixed values to transmit or block fault data; or a choice of ways to set up the fault condition on the node in the first place. The problem with any situation like this is that the selected choice can be wrong, leading to an inconsistent logic value on a particular node. If this should occur it is necessary to backtrack to the choice point, select another possibility, and try again. The process becomes complex if choice points are cascaded.

Exercise 3.2, based on Figure 3.14, illustrates the nature of the problems associated with choice possibilities.

Exercise 3.2

This exercise refers to the logic circuit shown in Figure 3.14 and illustrates something of the problems that can be caused when there is a choice of possibilities for propagating the effect of a stuck-at fault to a primary output. The exercise, primarily, is to generate a test for a stuck-at-1 fault on U6.12.

Note: The convention for identifying a particular location on a device is to first write the device identifier and then the pin number on the device. These two items are separated by a period. U6.12 therefore

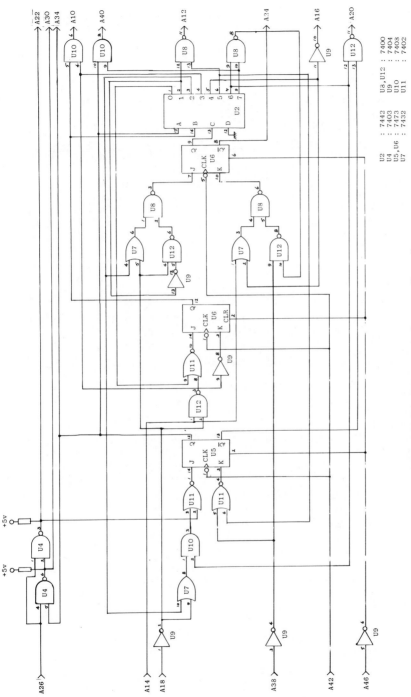

FIGURE 3.14 Example 8 (Courtesy Genrad)

61

SN7442 BCD-TO-DECIMAL DECODER (1-OUT-of-10)

NO.	BCD INPUT				DECIMAL OUTPUT									
	D	C	B	A	0	1	2	3	4	5	6	7	8	9
0	L	L	L	L	L	H	H	H	H	H	H	H	H	H
1	L	L	L	H	H	L	H	H	H	H	H	H	H	H
2	L	L	H	L	H	H	L	H	H	H	H	H	H	H
3	L	L	H	H	H	H	H	L	H	H	H	H	H	H
4	L	H	L	L	H	H	H	H	L	H	H	H	H	H
5	L	H	L	H	H	H	H	H	H	L	H	H	H	H
6	L	H	H	L	H	H	H	H	H	H	L	H	H	H
7	L	H	H	H	H	H	H	H	H	H	H	L	H	H
8	H	L	L	L	H	H	H	H	H	H	H	H	L	H
9	H	L	L	H	H	H	H	H	H	H	H	H	H	L
INVALID	H	L	H	L	H	H	H	H	H	H	H	H	H	H
	H	L	H	H	H	H	H	H	H	H	H	H	H	H
	H	H	L	L	H	H	H	H	H	H	H	H	H	H
	H	H	L	H	H	H	H	H	H	H	H	H	H	H
	H	H	H	L	H	H	H	H	H	H	H	H	H	H
	H	H	H	H	H	H	H	H	H	H	H	H	H	H

H = high level, L = low level

SN7473 DUAL J-K FLIP-FLOP (-VE CLOCK)

INPUTS				OUTPUT	
CLEAR	CLOCK	J	K	Q	$\bar{Q}$
L	X	X	X	L	H
H	⌐_	L	L	Q_O	$\bar{Q}_O$
H	⌐_	H	L	H	L
H	⌐_	L	H	L	H
H	⌐_	H	H	TOGGLE	

FIGURE 3.15 SN7442 and SN7473 Truth Tables

means device U6, pin 12 and is, in fact, the Q output of the J-K flip-flop located at position U6.

The word "node" is used to indicate a connection from one point to another. If there is no symbolic name associated with a node, it is usually referred to by the same name as the device and pin number from which it originates, not the device and pin number of the device on which it terminates. In this way, a node from a device that drives more than one loading device is named unambiguously.

Part 1. How many ways are there for setting up the initial value of 0 on U6.12?

Part 2. How many single paths are there from the source of the fault, U6.12, to the primary outputs?

Part 3. How many ways are there for transmitting the effect of the fault through the U2 decoder device?

Part 4. How many ways are there for blocking the effect of the fault through U2?

Part 5. Generate a test for U6.12 s-a-1.

Exercise 3.2: Outline Solution

Part 1. To initialize U6.12 to 0, there are two possibilities:
  i) By using the clock input. Set U6.14 (J input) to 0, U6.3 (K input) to 1, and clock negative edge) on A42 (SYSCLK).
  ii) By using the reset facility. Set A46 (RESET) to 1 and release to 0.

Parts 2–4. Study the table overleaf for the SN7442 decoder, device U2.

This table shows the various input patterns that can be set up on the input side of the decoder and the corresponding output selected to go to 0. For any particular input combination, we can assess the change on the output values if the value on the B input (U2.14) changes $0 \rightarrow 1$. This particular input connects directly to the source of the fault, U6.12.

Input values on				Output selected	Output changes
U2.12	U2.13	U2.14	U2.15	low	if input B(U2.14)
D	C	B	A		changes 0→1
0	0	0	0	0 (U2.1)	0(0→1), 2(1→0)
0	0	0	1	1 (U2.2)	1(0→1), 3(1→0)
0	0	1	0	2 (U2.3)	No changes
0	0	1	1	3 (U2.4)	No changes
0	1	0	0	4 (U2.5)	4(0→1), 6(1→0)
0	1	0	1	5 (U2.6)	5(0→1), 7(1→0)
0	1	1	0	6 (U2.7)	No changes
0	1	1	1	7 (U2.8)	No changes

The expected output changes are shown to the right of the table and, depending on the input combination selected (from a choice of four combinations), the effect of the fault can be propagated simultaneously to any two of the outputs, taken in the pairs shown and covering all outputs collectively.

The answers to Parts 2, 3, and 4, are therefore as follows:

Part 2. There are nominally nine single paths—eight passing through the U2 decoder and thence to various outputs or, in some cases, creating a closed feedback loop, e.g.

U6.12 → U2.14 → U2.1 → U11.9 → U6.14 → U6.12

The ninth path is the direct path from U6 to U10 and out on A10.

Note also that all paths through U2 contain routes back into the logic as well as to the primary outputs. This can mean possible interference through reconvergence with values that have been fixed either to initialize the U6 flip-flop or to set up the other input values to the U2 decoder.

Part 3. As the table shows, there are four input combinations that will allow the transmission of the fault effect on U2.14 through the U2 device to its outputs.

Part 4. There is no input combination that blocks the fault-effect transmission through U2. Because of the direct connection between the source of the fault, U6.12, and U2.14, it is not possible to set U2.14 to 1 and still have U6.12 nominally at 0.

In general, it is often necessary to block a fault-effect transmission. There will probably be alternative ways of doing this.

Part 5. Because there are so many paths from the source of the fault to the primary outputs, there are, potentially at least, many tests for the fault. The obvious candidate for a first attempt is the direct path from U6.12 to primary output A10 via U10. This path will exist if the other input to U10, U10.4, coming from U2.4, can be fixed at 1. Since the effect of the fault cannot be blocked through U2, we must select an input combination that leaves U2.4 at 1 even if U2.14 changes. This restricts the choice of input combinations on U2 to three, i.e., $D = 0, C = 0, B = 0, A = 0$ or $D = 0, C = 0, B = 0, A = 1$, or $D = 0, C = 1, B = 0, A = 1$.

The next stage is to select one of these three choices and attempt to establish the values, working back through the circuit. The choice is probably arbitrary at this stage but, if inconsistencies (clashes of values) occur, we need to remember which choice we made so that we can back-track to U2 and select another choice. The remainder of this exercise is left to the reader.

Completion of Exercise 3.2 should result in a number of discoveries. First, the possibility of choice arises from four main areas:

(i) The selection of initial values to excite the fault condition;

(ii) The number of paths along which a fault effect can be transmitted;

(iii) The way in which a fault effect can be transmitted through a multi-output device;

(iv) The way in which a fault effect can be blocked through a device.

When a choice of possibilities exists, there is usually an obvious first candidate. Troubles arise when this choice turns out to be incorrect and the need arises to go back and re-select. It is necessary to remember what the original set of choices were, which have been tried already, and which remain. Complications arise when there is a layer of choices, that is, when the choices become cascaded. For a complex board with many choice situations, it may become impossible to try all possibilities, with the result that a test is not generated for a particular fault even though one probably exists.

The second discovery arising from Exercise 3.2 is that some way of marking the circuit diagram with logic 0's and 1's is required so that values can be changed when re-selection from a set of choices takes place. One possibility, which works well in practice, is to tape or glue the circuit diagram to a backing sheet of thick cardboard and use map pins of different color to denote logic 1's and logic 0's. The pins can be changed many times and allow the status of the circuit to be seen "at a glance," as it were.

The third discovery from the exercise is that it is necessary to have data sheets for the devices on the board and, furthermore, that these sheets must be understood. This may seem obvious but there can be a tendency to guess the behavior of a device if a similar device has been studied recently. This is particularly true of flip-flops, for example. Even within the series 74 range of TTL-JK flip-flops, there are many varieties and it is important to know exactly how a particular device behaves—which edge it clocks on and which level causes preset or clear functions.

Finally, the exercise should have demonstrated the need to be orderly. This means working from an overall test plan and, even for individual tests, there may be a need to record choices tried and those so far untried. In the end, the amount of information recorded on paper and the amount memorized is a personal affair. If it ever becomes necessary to revise the test program, some record of the reasons for and against various decisions becomes invaluable.

3.6 Comments on an Overall Test Strategy

To conclude these two chapters on the theory of test-pattern generation, some observations of a general nature will be made on the strategy for developing a test plan. There are two fundamental approaches to writing a test program. The first, or functional approach, is based on the desire to show that the circuit performs the correct function. The second, called the structural approach, is based on the desire to show that the circuit contains no faults; in this respect, the structural approach is often called the "fault-oriented" or "fault-driven" approach. In developing a test plan for a logic circuit therefore, the test programmer is faced with the critical question: should the main approach be functional or structural? The functional approach is usually quick to formulate and write down but may miss certain faults, whereas the structural approach gets right to the heart of the matter, that is, to the faults themselves, but can often become protracted for the various reasons illustrated in the previous examples. There is no "best solution" as such—each board is different, poses different problems, and requires different specific solutions. What appears to work well in practice, however, is the strategy outlined in Figure 3.16.

Figure 3.16 suggests that, initially, the testing strategy should be functional. Unfortunately, the overall or "global" function of the

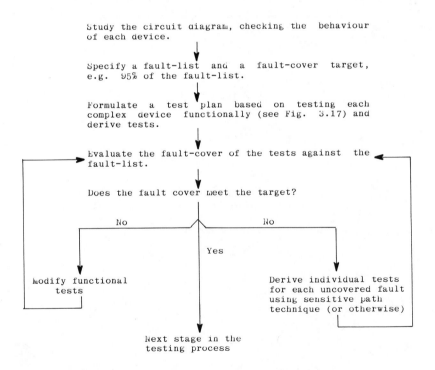

FIGURE 3.16 Practical Strategy for Developing a Test Program

circuit is usually not known. For instance, what is the overall function of Example 8 (Figure 3.14)?

What is known, however, is the individual or "local" function of each of the devices in the circuit. If it can be shown that these devices are functioning correctly then, to a large extent, the global function will also be checked. This latter statement must be true because, for each device tested, we will need both to control and observe through other intermediate devices and their interconnections. If all major devices are tested in this way, then most, if not all interconnections will also be tested. These observations are summarized in Figure 3.17.

The difficulty with this approach is that sometimes it becomes difficult, if not impossible, to put each device fully through its paces. There is also the problem of specifying a functional test for each device. Nevertheless, it is usually possible to write a number of tests fairly quickly and, at some stage, evaluate their effectiveness in detecting

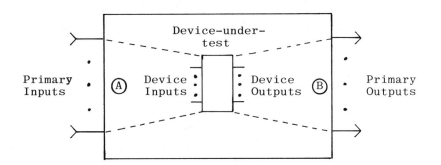

Region A: In this region, fixed (fault-free) values are established from the primary inputs to the device inputs via the intermediate predecessor devices to establish the test input requirements.

Region B: In this region, sensitive paths are set up from the device ouputs to the primary outputs, via the intermediate successor devices, to allow observation of the output responses.

FIGURE 3.17 Controllability and Observability

actual faults. In practice, a high proportion of real faults, or simulated stuck-at-1, stuck-at-0 faults, will be detected by these tests, thereby reducing the list of outstanding faults to a manageable size. The flow-chart in Figure 3.16 suggests that a choice of strategy now exists: either return and modify the functional tests to try and cover yet more faults, or generate extra individual tests using sensitive path techniques for each of the remaining uncovered faults. Learning where the trade-off is between these options is something that comes only with experience. If the initial fault cover is as high as 70% to 80% of the fault cover target figure with a local functional strategy, then the tendency is to find that the uncovered faults are fairly closely related to each other. In other words, if a test is generated for one of them, it may well turn out to be a test for some of the others.

A useful exercise at this stage is to return to the logic circuit in Figure 3.14 and write down a test plan. As before, there is a worked solution to this exercise but be warned—there is very seldom a unique answer to these problems.

Exercise 3.3

This exercise is an example of how a test plan for a logic circuit might be constructed. The circuit in question is shown in Figure 3.14.

It is not necessary to consider the tests in detail. The end product should be a sequence of steps or a flow-chart outlining the sequence of events, i.e., the order of testing the devices and how each device will be tested. Assume that, at this stage, the only information available is the circuit diagram and the data sheets for the logic devices (Figure 3.15).

Exercise 3.3: Outline Solution

Note: This is one of many solutions to the problem.
    Test plan based on local functional strategy:

Step 1 Initialize circuit using A46 (RESET) input to initialize flip-flops and set all other primary inputs low (logic 0). Release the RESET input.

Comment: Selecting all other inputs to be low is quite arbitrary. The point about initialization is that the value of every node in the circuit should be known, i.e., primary input, primary output, and internal. This may require selection of primary input values on an individual basis.

Step 2 Test the clocked operation of the first flip-flop, device U5.

Comment: The somewhat complex feedback structure of this circuit does not suggest a natural "device-to-test-first." As a general maxim, test the easy sections before tackling the difficult sections.

As far as the flip-flop is concerned, at least four tests would seem useful, i.e., the four different sets of values on the J and K inputs, each followed by a negative clock-edge on the system clock A42 (SYSCLK).

Note that one of these tests, J = K = 0, would be a "no-change-expected" test and could be repeated, once for Q = 1, $\overline{Q}$ = 0 and the second time for Q = 0, $\overline{Q}$ = 1. Similarly, the "toggle" test (J = K = 1) could be repeated to demonstrate a toggle in both directions.

Observation of the result of each test can take place on A20 (SACK) for the $\overline{Q}$ output *and* A40 (ENA) for the Q output. It is important to observe and test both outputs of the flip-flop. Note also that it may not be possible to set up input-control and output-observation conditions for all the tests.

Step 3 Test the middle flip-flop, device U6, using the same functional tests as Step 2.

Comment: Again, the nature of the feedback in the circuit may make it impossible to carry out all the tests. The answer here is to carry out as many as possible.

Observation of the Q output of this flip-flop should be, as far as possible, on an output not yet used. As we have seen, there are many possibilities for observing this output.

Step 4 Test the rightmost flip-flop, device U6, using the same functional tests as for Step 2.

Comment: The $\overline{Q}$ output of this flip-flop is directly observable on A24 (QBAR). The Q output is only observable via the decoder, U2.

Step 5 Test the SN 7442 decoder, device U2.

Comment: A functional test for this decoder consists of the eight possible input combinations on the A, B, and C inputs (input D is tied down to Ov). Some of these combinations will have occurred in the previous tests on the flip-flops and a number of options exist:

   i) Ignore the fact that some input combinations have already been applied and try to test the decoder using all eight combinations;

  ii) Go back to the previous steps and identify those particular combinations that have been applied to the decoder during the course of the tests on the flip-flops. Maximize the fault-cover of each test by increasing the number of sensitive paths that come from the decoder outputs. Finally, add extra tests only for the combinations that have not yet occurred;

 iii) The third option is to start the whole test plan again, this time starting with the decoder rather than with the flip-flops. This is done on the grounds that, in testing the decoder, all three of the flip-flops will have to change their states. In effect, therefore, a complete functional test on the decoder will probably exercise most of the remainder of the circuit.

It is not uncommon to revise a test plan as understanding of how the circuit functions as a whole increases. The main thing really is that there should be an initial test plan to give some structure to the progression through the process of generating tests.

Step 6 Evaluate the full fault-cover of the tests generated so far.

Comment: Assuming that we stay with the original test plan as defined by the previous steps, we have now reached a stage when it would be worth while evaluating the full fault cover of all the tests. What we will find is that not only have we tested the three flip-flops and the decoder, but we must also have tested a fair number of faults on the NAND and NOR gates because these gates have been used as fault transmitters.

Step 7 Complete tests up to target fault cover, either by modifying earlier tests or by adding extra tests aimed specifically at the remaining uncovered faults.

Comment: It is usually safer and, in the long term, quicker not to return to earlier tests and attempt to increase their fault-cover, but to write new tests for each of the uncovered faults. The maxim here is ''leave well enough alone'' if the earlier tests do the job they were intended to do. Sometimes, all the repercussions of a modification to an earlier test are not immediately apparent and much time can be wasted. In other words, beware the ''perfection at infinity'' syndrome.

As a final exercise, Exercise 3.4 is based on a different circuit (Figure 3.18) and illustrates several points of specific detail not brought out by the earlier examples.

Exercise 3.4

Formulate an overall test plan for the circuit shown in Figure 3.18 (Example 9). The SN 7405 is an open-collector hex inverter device : the behaviors of the SN74157, and SN74161 devices are summarized in Figures 3.19, 3.20, and 3.21.

Exercise 3.4: Outline Solution

General observations

The circuit consists essentially of:

    i) A 2-stage shift-register with feed-back (device IC1);

    ii) A 4-bit binary up-counter with pre-load facilities (device IC2) whose outputs pass directly to the ''B'' inputs of a multiplexer (device IC3);

    (iii) Two cross-coupled NAND gate latches (device IC4);

    iv) Some 2-input NAND gates (device IC6) and open-collector

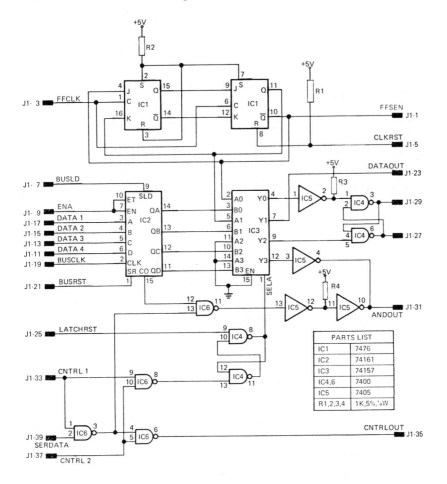

FIGURE 3.18 Example 9 (Courtesy Computer Automation)

inverters (device IC5), some of which have pull-up resistors on the board.

Note the following special features:

i) The outputs IC5.4 and IC5.10 are wired together to form the J1-31 primary output. This output is the logical AND between IC5.4 and IC5.10 (i.e., wiring together the outputs of open-collector inverters results in a wired-AND for positive logic).

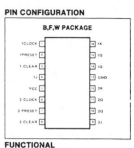

DUAL J-K FLIP-FLOP   **54/7476**

SPEED/PACKAGE AVAILABILITY

54	F,W	74	B,F
54H	F,W	74H	B,F
54LS	F,W	74LS	B,F

PIN CONFIGURATION

B,F,W PACKAGE

1CLOCK	1	16 1K
1PRESET	2	15 1Q
1 CLEAR	3	14 1Q̄
1J	4	13 GND
VCC	5	12 2K
2 CLOCK	6	11 2Q
2 PRESET	7	10 2Q̄
2 CLEAR	8	9 2J

DESCRIPTION

This monolithic dual J-K flip-flop features individual J, K, clock, and asynchronous preset and clear inputs to each flip-flop. The preset or clear inputs, when low, set or reset the outputs regardless of the levels at the other inputs. When preset and clear inputs are inactive (high), a high level at the clock input enables the J and K inputs and data will be accepted. The logic levels at the J and K inputs may be allowed to change when the clock pulse is high and the bistable will perform according to the function table as long as minimum setup and hold times are observed. Input data is transferred to the outputs on the negative-going edge of the clock pulse.

FUNCTION TABLE (Each Flip Flop)

54/74,54/74H						
INPUTS					**OUTPUTS**	
PRESET	CLEAR	CLOCK	J	K	Q	Q̄
L	H	X	X	X	H	L
H	L	X	X	X	L	H
L	L	X	X	X	H*	H*
H	H	⊓	L	L	Q₀	Q̄₀
H	H	⊓	H	L	H	L
H	H	⊓	L	H	L	H
H	H	⊓	H	H	TOGGLE	

54/74LS						
INPUTS					**OUTPUTS**	
PRESET	CLEAR	CLOCK	J	K	Q	Q̄
L	H	X	X	X	H	L
H	L	X	X	X	L	H
L	L	X	X	X	H*	H*
H	H	↓	L	L	Q₀	Q̄₀
H	H	↓	H	L	H	L
H	H	↓	L	H	L	H
H	H	↓	H	H	TOGGLE	
H	H	H	X	X	Q₀	Q̄₀

H = high level (steady state)
L = low level (steady state)
X = irrelevant
↓ = transition from high to low level
Q₀ = the level of Q before the indicated steady-state input conditions were established.
TOGGLE: Each output changes to the complement of its previous level on each ↓ clock transition.
*This configuration is nonstable, that is, it will not persist when preset and clear inputs return to their inactive (high) level.

FUNCTIONAL BLOCK DIAGRAM

FIGURE 3.19 SN7476 (Courtesy Signetics)

ii) There is no pull-up resistor on either IC5.4 or IC5.10. This pull-up will have to be supplied, either by the tester or by means of a special fixture on the board.

Finally, a possible overall test strategy could be based on the hybrid approach of testing each main device in a functional manner and then "topping-up" with specific tests for any uncovered faults. The following test plan is based on this approach.

Phase 1 Initialize the circuit.
   Comment: The only real problem here is that the first stage of the 2-stage shift register is not directly accessible for initialization. One solution is to initialize the second stage (using the J1–5 input) and, holding this second stage down, to clock the known outputs of the second stage (IC1.11 and IC1.10)

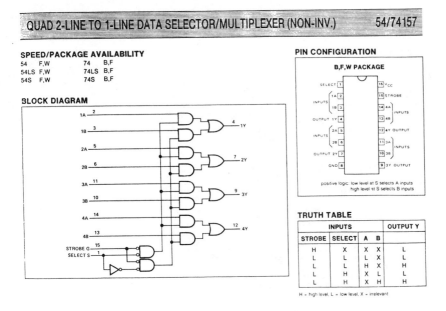

FIGURE 3.20 SN74157 (Courtesy Signetics)

through the first stage (i.e., on to IC1.15 and IC1.14) with a negative edge on J1–3.

As for the other devices, IC2 is directly initializable using the reset input, J1–21; IC3 can be selected to pass the counter outputs (IC2.11–14) straight through by setting IC3.1 to logic 1 (achieved by setting J1–25 to 0). All other values in the circuit, primary or otherwise, can be defined unambiguously.

Phase 2 Test 2-stage shift register (device IC1)
Comment: The outputs of this shift register are observable either directly (on J1–1) or via IC3 (J1–29 and J1–23). Observation via IC3 will require IC3.1 at 0 to select "A" inputs through to "Y" output.

The test on the shift register itself will consist of five clock pulses to take the register through each of its four possible states and return it to the start state. There should also be a separate test on the clock input (J1–3) to ensure that it is not s-a-1 or s-a-0.

SYNCHRONOUS 4-BIT BINARY COUNTER   54/74161

SPEED/PACKAGE AVAILABILITY

54	F,W	74	B,F
54LS	F,W	74LS	B,F

PIN CONFIGURATION

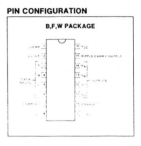

B,F,W PACKAGE

DESCRIPTION

This synchronous presettable binary counter features an internal carry look-ahead for applications in high-speed counting designs. Synchronous operation is provided by having all flip-flops clocked simultaneously so that the outputs change coincident with each other when so instructed by the count-enable inputs and internal gating. This mode of operation eliminates the output counting spikes which are normally associated with asynchronous (ripple clock) counters. A buffered clock input triggers the four flip-flops on the rising (positive-going) edge of the clock input waveform.

This counter is fully programmable; that is, the outputs may be preset to either level. As presetting is synchronous, setting up a low level at the load input disables the counter and causes the outputs to agree with the setup data after the next clock pulse regardless of the levels of the enable inputs. The clear function for the 54/74LS161 is asynchronous and a low level at the clear input sets all four of the flip-flop outputs low regardless of the levels of clock, load or enable inputs.

The carry look-ahead circuitry provides for cascading counters for n-bit synchronous applications without additional gating. Instrumental in accomplishing this function are two count-enable inputs and a ripple carry output. Both count-enable inputs (P and T) must be high to count, and input T is fed forward to enable the ripple carry output. The ripple carry output thus enabled will produce a high-level output pulse with a duration approximately equal to the high-level portion of the Q_A output. This high-level overflow ripple carry pulse can be used to enable successive cascaded stages. Transitions at the enable P or T inputs are allowed regardless of the level of the clock input.

The 54/74LS161 features a fully independent clock circuit. Changes made to control inputs (enable P or T, load or clear) that will modify the operating mode have no effect until clocking occurs. The function of the counter (whether enabled, disabled, loading or counting) will be dictated solely by the conditions meeting the stable setup and hold times.

BLOCK DIAGRAMS

54/74

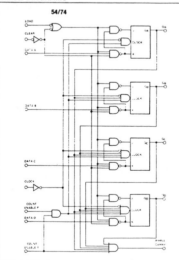

TYPICAL CLEAR, PRESET, COUNT, AND INHIBIT SEQUENCES

Illustrated below is the following sequence:

1. Clear outputs to zero
2. Preset to binary twelve
3. Count to thirteen, fourteen, fifteen, zero, one, and two
4. Inhibit

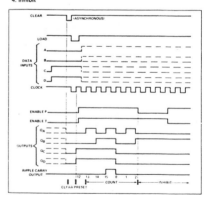

FIGURE 3.21 SN74161 (Courtesy Signetics)

Phase 3 Test the synchronous 4-bit counter (IC2)

Comment: The main observation on the outputs of this counter is via the IC3 multiplexer. The select input therefore, IC3.1, must be set to 1. The tests themselves should include a test for:

i) The reset condition;

ii) Normal clocked up-count (release reset input and clock 16 times);

iii) The preload facility (preload all-0's, all-1's, chequerboard 0101, and inverse chequerboard 1010);

iv) The count-enable (J1–9) function (check that the counter does not count when J1–9 is low);

v) The carry-out line (IC2.15) when count reaches all-1's (observable via a sensitive path through IC6.12 → IC6.11 → IC5.13 → IC5.12 → IC5.11 → IC5.10 → J1–31).

Phase 4 Test the multiplexer (device IC3)

Comment: This device has probably been tested by the earlier phases but a simple test can be included here by preloading the counter with 1's and selecting the "A" inputs (which must contain at least three 0's) and then the "B" inputs (four 1's). This will cover both stuck-at and bridging faults on the inputs to this device.

Phase 5 Generate specific tests for remaining NAND gates (device IC6) and latches (device IC4)

Comment: Again, it is likely that some stuck-at-faults on the inputs and outputs of these gates will have been covered by previous tests. This phase can either 'be omitted until the fault-cover of the earlier tests has been determined, or simple tests included for each gate based on:

i) Setting up the 10, 01, and 11 inputs;

ii) Establishing a sensitive path from the output to a primary output.

Note: A 2-input NAND gate is fully tested for stuck-at-0 and stuck-at-1 faults on both inputs and output by the three input patterns of 10, 01, and 11.

3.7 Final Comment on TPG : The D-Algorithm

Chapters 2 and 3 have aimed at giving an intuitive understanding of the problems involved with test-pattern generation. The subject is one that has attracted much attention from members of the academic and industrial communities and there have been many attempts to formalize the theory in order to produce a general-purpose algorithm that can be used as a basis for algorithmic test-pattern generation (ATPG).

Unfortunately, many of the algorithms reported in the literature suffer from one or more of the following limitations.

(i) They are only applicable to acyclic circuits (circuits without stored-state devices or global feedback).

(ii) They are only applicable to sequential circuits which are described by a formal state-diagram or state-table. In practice such information rarely exists and is difficult to generate.

(iii) Many of the algorithms are formulated with very little regard for their applicability to "real-life" circuits, i.e., to actual number of primary inputs/primary outputs, circuit complexity, presence of non-standard devices such as monostables or delay elements, devices with tristate or open-collector facilities, etc.

(iv) The storage and run-time requirements of a programmed implementation have not been considered. Often these requirements become the limiting factor for a practical ATPG.

Despite these criticisms, however, some algorithms have enjoyed a limited success; the best known of these is the "D-algorithm," first proposed by research workers at IBM in 1966. The algorithm is really a formalization of the concepts of fault-effect propagation along single or multiple sensitive paths and, initially at least, was only applicable to acyclic circuits. There have been some attempts to extend the theory into the sequential-circuit domain, but success has been more limited here. Because of its importance, however, the algorithm is described in the appendix to this book. It is not necessary to read the appendix before reading the remaining chapters of the book, but reading it is recommended for those wishing to pursue TPG further. Most, if not all successful algorithms for generating test patterns are variations of the D-algorithm.

Chapter 4
Test-Pattern Evaluation

The first acceptance criterion for a set of test patterns is the fault cover expressed as a percentage of faults in the fault list. This process of test-pattern evaluation is the subject of Chapter 4.

In practice, there are two techniques used to determine this percentage. The first is to insert actual faults onto the pins of the devices on a known-good-board and run the set of patterns. This is called ''physical fault insertion.'' The second is to simulate the response of the board to the test-pattern inputs, and to do this initially without a fault in order to obtain fault-free reference data, and then in the presence of faults.

4.1 Evaluation by Physical Fault Insertion

The general procedure for this method is to set up the board on the tester and run the test program in an endless loop. Individual faults can then be selected from the fault list and inserted onto the appropriate node of the circuit. The test program will either detect the fault or the board will be passed. In this way, therefore, uncovered faults will be isolated.

The main problem with this technique is that faults can only be inserted if they do not cause irreversible damage to the board. For standard TTL devices, this amounts to stuck-at-0 faults only.* It is not

*High power TTL and Schottky TTL devices can only be driven low for a period of 1 second or so.

advisable to tie the output of a TTL gate up to +5v for any length of time, either through a resistor or not, because this will, in due course, cause damage to the lower totem-pole transistor. (Some testers have facilities for pulse-driving a TTL output for a period long enough to simulate a stuck-at-1 fault but not long enough to damage the transistor.)

Other technologies, such as CMOS and ECL, allow both levels of stuck-at fault insertion (See Figure 4.1).

Another problem is that it is not possible to insert certain classes of fault, such as those that introduce propagation delay faults. In practice these are an important class of fault for high-speed circuits, as noted in earlier chapters.

Altogether therefore, although physical fault insertion is a simple way of evaluating the performance of a set of tests, the results are only partial. For this reason, the alternative method based on the use of a logic simulator is used almost universally by test programmers.

4.2 Evaluation by Simulation

The simulation process consists of defining a model of the system and then exercising this model with defined sets of input stimuli in order to produce a record of this signal-versus-time behavior. A logic simulator is a computer program capable of implementing this process. The two principal use of simulators therefore, are:

Logic family	V_L	V_H	V_T	Physical fault techniques S-a-0	S-a-1
TTL	0v	+5v	+1.4v	Direct short to 0v*	Not possible unless pulsed
CMOS	0v	+12v	+5v	Direct short to 0v	Direct short to +12v
ECL	-2.1v	-0.6v	-1.4v	Through pull-up resistor to -2.1v	Through pull-up resistor to -0.6v

*But not high-power and Schottky TTL : see text

FIGURE 4.1 Simulating Stuck-At Faults

(i) to verify, or otherwise, the function of the initial design by exercising the model in an assumed fault-free mode of operation, and

(ii) to evaluate the effectiveness of engineer-specified, pseudo-random, or algorithmically-generated test patterns.

In this chapter we will be more concerned with (ii) above than with (i) and in this section we will summarize the main features of a simulator.

4.2.1 Modelling Techniques

The two main approaches to modeling are the ''compiled-code'' and ''table-driven'' approaches. In the compiled-code approach, the model is formed using the input statements of the source language. These statements are compiled as a set of program segments (subroutines or integer functions) that correspond to the logical functions.

As a simple example of this, the following integer function is a possible FORTRAN listing for a 2-input AND gate:

```
INTEGER FUNCTION AND2 (I,J)
AND2 = 0
IF ((I.EQ.1).AND.(J.EQ.1)) AND2 = 1
RETURN
END
```

A call to this function would now be made in the usual way, e.g., for a 2-input AND gate with inputs x_1, x_2 and output z_1, the calling statement in the master segment would be

Z1 = AND2 (X1,X2)

In the table-driven approach, use is made of list-processing data structures to set up lists (tables) corresponding to the various attributes of the circuit. These lists include:

(i) a circuit description table detailing the topology of the circuit;

(ii) an initial input-value table containing the input stimuli;

(iii) signal value tables containing the Boolean value derived for each primary input, primary output, or internal connection carrying a logic signal;

(iv) a precedence or timing table detailing the order in which the various elements within the circuit are to be analyzed;

(v) a function description table containing data about the logical operation of the various elements that make up the circuit.

Note:

(i) The contents of these tables are either specified by the designer as initial conditions, are derived during the simulation process, or are already available as standard library tables.

(ii) The compiled-code and table-driven approaches can be combined by making the pointers and data in the function description table the name and arguments of the appropriate program segments.

4.2.2 Other Considerations

The following factors are other considerations to be taken into account when developing or assessing the usefulness of digital simulators.

(i) 2-value, 3-value, or n-value

The number of values of a simulator relates to the different assignments that can be made to the logic signal lines. A 2-value simulator is the basic system, the two values being 0 and 1. A 3-value simulator includes an additional symbol X to denote an unknown status on the line. Figure 4.2 shows the line values that would be generated for a 2-input AND gate whose inputs are both changing, one going 1 to 0 and the other 0 to 1. The 3-value output indicates the possibility of a static-0 hazard.

The 2-value simulator is normally only used for logic verification, whereas the 3-value simulator allows hazard and race conditions to be detected (as indicated in Fig. 4.2(b)).

Some simulators increase the number of values beyond 3. One example is a 6-valued simulator using the values 0, 1, X as before, plus U (signal value changing from 0 to 1), D (signal value changing from 1

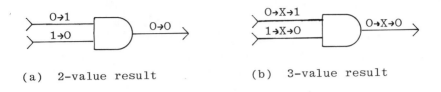

(a)   2-value result   (b)   3-value result

FIGURE 4.2 2-Value and 3-Value Simulation

to 0), and H or $\overline{H}$ to denote the presence of a potential static-1 or static-0 hazard. The more detailed the number of values, the more detailed the race and hazard analysis.

(ii) Zero, unit, or assignable propagation delays and signal rise and fall times.

Obviously the accuracy of the simulator results increases according as the precision of the element propagation delays and signal rise and fall times is increased. The options available are usually to assume instantaneous response (zero delay) or to assign a unit fixed delay to each element with zero rise and fall times, or to allow user-specified delays and rise and fall times to be associated with individual elements. It is also possible to adopt probabilistic approaches using a pseudo-random number generator to act as a selector address in a look-up table containing delay or other statistical data. This is the so-called ''Monte-Carlo'' approach.

The price paid for increased precision in assigning delay and signal transition times is that the simulator development becomes more complex and the execution time is increased. Both these factors increase the cost.

(iii) Mechanism for incrementing time

The progression of time through a simulation run can either be in ''fixed increments'' (synchronous) or ''event-driven'' (asynchronous).

In the fixed-increment simulator, the occurrence and assignment of values is tied to a fixed time interval, possibly the time between consecutive clock pulses. This approach has the advantage of simplicity but may miss inter-clock hazard conditions. It is also possible that there could be long periods of inactivity.

In the event-driven simulator, a table of next-events is held in a push-down stack organized on a strict chronological basis. The top item is the next chronological event and this is pulled off the top of the stack and initiates the next phase of simulation.

An event-driven simulator is usually more accurate than the fixed-increment simulator but has a more complex structure.

Associated with this aspect of simulators is the "selective trace" capability. Generally, when the Boolean values of logic-carrying signal lines are updated, it is usual to update all lines irrespective of whether they are affected or not. The selective trace facility identifies only those portions of the network that are affected by the input stimulus change and restricts the attention of the simulator accordingly. This has the overall effect of reducing the operational time of the simulator and hence the running costs.

(iv) Macro facilities

Some simulators allow the user to define complex elements (such as shift-registers) in terms of the basic elements in the library (J-K flip-flops, say). Once defined, these descriptions, called "macros," pass into the library as new named elements and can be called simply by name. This is obviously a powerful strategy for widening the application of the simulator and, in essence, allows the designer to look at the system behavior at either basic gate level or at a higher level.

4.3 Application of Simulators to Testing Problems

The main applications of a logic simulator to testing problems are:

   (i) to provide assistance in the processes of test-pattern generation;
   (ii) to determine individual and composite fault covers;
   (iii) to produce test-program support files and other documentation.

In reality therefore, the simulator is but one part of a total software system. Figure 4.3 shows the main elements of such a system. These elements are discussed in the following subsections.

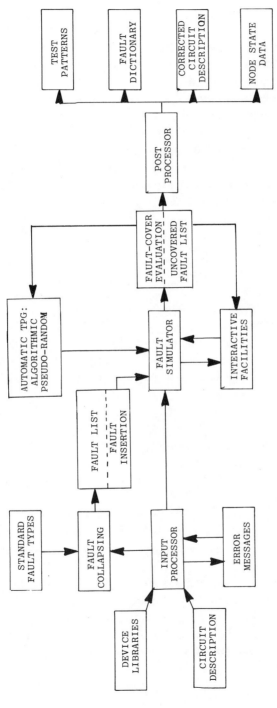

FIGURE 4.3 Simulator-Based System

4.3.1 Input Processor

The function of the input processor is to use the circuit description and device libraries to prepare a suitable model for the simulator. The circuit description consists of a list of the circuit devices together with the interconnects between them. This description is generally known as an "image." Usually, the devices are identified both by type (SN7473, say) and location identifier (U12, say). An interconnect between one device (U12, pin 3, say) and another (U8, pin 6, say) is either assigned a symbolic name or is described by the identifier and pin number of the driving device (not the driven). For example, if U12, pin 3 is a device output which drives into U8, pin 6, then the connectivity between these devices could be listed as:

U12, . . . , 3 = U12.3
U8, . . . , 6 = U12.3

Note that assigning symbolic names according to the driving device takes care of fanout possibilities. Effectively, all driven devices (fanout branches) will be connected to the same source (fanout trunk).

The device libraries consist of a standard-device library and, possibly, a user-specified library. The standard-device library contains models for the behavior of a range of standard commercial or in-house devices. These models may be in the form of:

(i) a truth-table for primitive gates and other combinational devices such as multiplexers or decoders;
(ii) a transition table for clocked devices such as flip-flops, counters, shift-registers, etc;
(iii) an equivalent primitive-gate model (usually NAND gate) for more complex devices such as PLAs or other memory devices;
(iv) a higher-level description for other LSI devices.

Examples of (i)–(iii) above are common and exist throughout the earlier chapters of the book. To illustrate (iv) however, the following is a high-level description of a logic circuit containing four primary inputs, a,b,c,d; two primary outputs, $z1$ and $z2$; and one internal state variable, y. The description is based on the Functional Definition Language used by Bell Laboratories. In this description, key words of the language are shown in upper case whereas arguments are shown in lower case.

```
 1 GPF : seq;
 2 INPUTS : a,b,c,d;
 3 OUTPUTS : z1,z2;
 4 BOOLEAN : y
 5 DEF
 6   IF a & b
 7   THEN z1 = 0; z2 = 0; y = 0
 8   ELSE
 9     IF a&b'&y'
10     THEN y = 1; zl = cUd; z2 = 0
11     ELSE z1 = 1; z2 = 1
12     FI
13   FI
14 FED
```

In line 1, GPF stands for general-purpose function and defines "seq" as the name of the function. Lines 2 and 3 are self-explanatory. Line 4 defines the internal state variable y to be Boolean. Line 5 identifies the start of the definition. Between lines 6 and 11, the symbols &, ', U denote logical AND, INVERT, OR respectively. The behavior of the circuit is described between these lines. FI terminates an IF and FED terminates DEF. No delay has been specified so the delay assumed will be equal to the unit delay used as the standard by the simulator. In the absence of any change, the value of the Boolean variable y is equated to the last known value.

The board may also contain custom-built devices that are not in the standard-device library. In this case, the user can specify the behavior in a separate user-specified library using any of the techniques outlined earlier.

The input processor assembles the circuit description into a usable data-structure for the simulator, making use of the device libraries. In so doing, various checks can be made on the correctness of the circuit description, e.g., are there any device outputs not specified as primary outputs and not connected to any other device? Similarly, are there any device inputs not specified as primary inputs and not connected to any other device? Such outputs and inputs are called "floating" outputs and inputs and their presence usually indicates a typing error. Other "soft" errors may also be flagged. An example here could be a check on whether both inputs to a 2-input NAND gate are used. If only one input is used this may indicate an error or it may be because the gate is being used as an inverter, i.e., one input floating or tied high to V_{cc} via a resistor.

Whatever the problem, the user will need to make the appropriate correction and then, eventually, the input processor produces the simulator model. Note that the modelling process should be capable of handling tri-state and open-collector devices, bidirectional bus lines, and wired-OR/AND junctions.

4.3.2 Fault Insertion

The next stage is to produce a fault list for use either by the test-pattern generation procedures or by the fault-insertion procedure. The production of the fault list may be preceded by a fault-collapsing process (as described in Chapter 2). The faults themselves come from a list of standard fault types, such as:

(i)   single stuck-at-1, stuck-at-0 faults on primary inputs and device output nodes;
(ii) adjacent IC pin shorts (modelled either as wired-AND or wired-OR depending on the technology);
(iii) interconnect open-circuit;
(iv) loss of V_{cc} supply or 0v ground return.

The net result is a fault list against which the test patterns can be judged.

4.3.3 TPG and TPE

Test-pattern generation and evaluation is the central activity of the system; the general procedure is shown in Figure 4.4.

The TPG aids usually include automatic procedures based either on an algorithmic technique (such as the D-algorithm described in the

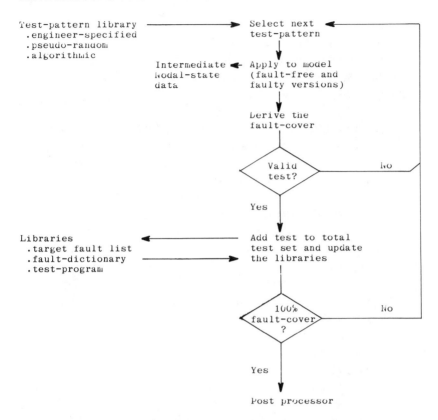

FIGURE 4.4 Use of Fault Simulator to Assist Test-Pattern Generation

appendix) or on pseudo-random pattern generation (discussed more fully in Chapter 6). Both these methods are limited in what they can achieve in terms of fault cover. As a consequence therefore, the third option is to allow direct input by the test programmers, usually interactively via a visual display unit (VDU).

The basic procedure is to select the next input stimulus, generate fault-free and faulty-state responses, derive fault cover and then determine whether the input is valid, i.e., whether the test covers faults so-far uncovered in the fault list. If the test is valid, it is added to the test-program library and the fault-dictionary and fault-list files updated accordingly. The procedure then returns to collect the next test if the fault-cover target has still not been reached.

A useful feature of this procedure is the display of nodal-state data, i.e., the current fault-free values on certain user-specified nodes. This information is essential for manual development of test patterns (cf. the map-pins technique discussed in the previous chapter—section 3.5).

Other useful facilities that can be included in an interactive environment are listed below.

(i) A test on the validity of a user-specified initialization procedure. Automatic provision of an initialization procedure can be very difficult depending on the complexity of the circuit and the number of simulation levels (0,1,X,etc.). Exercise 4.1 is designed to demonstrate some aspects of this problem.

(ii) Investigation of the behavior of the circuit in the presence of a fault with particular reference to the possibility of reintroducing race or hazard conditions. Such conditions, if they exist, can lead to indeterminate behavior thereby leading to difficulties of fault diagnosis.

(iii) The insertion of user-defined faults. Generally, almost any special fault can be added to the target fault list if the effect of the fault can be described logically. An example of this could be a bridging fault between two adjacent printed-circuit tracks which are physically routed close together but which do not go to adjacent device pins. This fault would not be modelled automatically by the fault-insertion procedure.

(iv) Variable timing parameters such as propagation delay, signal rise, and signal fall times. The simulator may adopt standard default values for these parameters, consistent with the technology modelled. The user, however, may wish to change the parameters to observe whether the test patterns are capable of detecting the change. This way of modeling the propagation-delay fault requires sophisticated timing control.

Exercise 4.1

Assuming the existence of a 4-state simulator $(0,1,X,\overline{X})$, consider the problem of initializing the circuit in Figure 4.5, given that all nodes start with an X or $\overline{X}$ status.

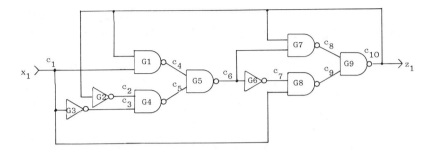

FIGURE 4.5 Example 10 (Exercise 4.1)

Exercise 4.1 : Outline Solution

The problem with initializing this circuit is that the value of the feedback variable c_{10} is unknown, for either $x_1 = 1$ or $x_1 = 0$. One solution is to assign a value X to c_{10} and then to assign other values throughout the circuit, based on the following 4-value truth tables for INVERTER and 2-input NAND gates.

Input	Output
0	1
1	0
X	$\overline{X}$
$\overline{X}$	X

Inverter

Inputs	0	1	X	$\overline{X}$
0	1	1	1	1
1	1	0	$\overline{X}$	X
X	1	$\overline{X}$	$\overline{X}$	1
$\overline{X}$	1	X	1	X

2-input NAND gate

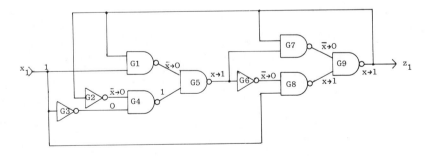

FIGURE 4.6 Example 10 Initialised

Figure 4.6 shows the result of this procedure, starting with an initial assignment of $x_1 = 1$ and $c_{10} = $ X. The order of evaluation is:

Connection	Value	
	First pass	Second pass
c_1	1	
c_{10}	$\overline{\text{X}}$	1
c_2	$\overline{\text{X}}$	0
c_3	0	
c_4	$\overline{\text{X}}$	0
c_5	1	
c_6	$\overline{\text{X}}$	1
c_7	$\overline{\text{X}}$	0
c_8	$\overline{\text{X}}$	0
c_9	X	1
c_{10}	1	

In this case, the second pass completed the assignment and produced no contradictory values. This is not always the case—for example, consider the oscillator circuit shown in Figure 4.7.

An initial control value of 0 on c_1 produces a stable set of internal nodal values, but a value of $c_1 = 1$ together with an initial value of $c_4 = $ X produces an oscillatory situation (as expected). The problems from the simulation viewpoint are:

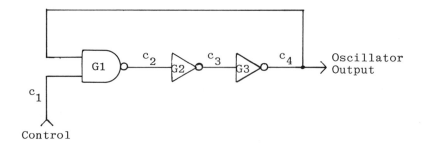

FIGURE 4.7 Oscillator Circuit

(i) to recognize the oscillatory result, and
(ii) to try to find another initial assignment that does not produce contradictory results.

4.3.4 Post Processor

When the user is satisfied with the performance of the test program, the data files created during the TPG/TPE loop are processed into a more permanent form by the post processor. The main files created are as follows:

(i) The test program
(ii) The fault dictionary and performance summary (fault-cover, undetected faults)
(iii) Circuit description
(iv) Node state record (fault-free values on internal nodes)

The fault dictionary is a list of input patterns together with the associated fault conditions that will cause the board to fail. This dictionary can form the basis for a fault-location strategy (see following chapter).

Once these files are available, the test program can be validated by applying it to a known-good-copy of the board. Unfortunately, it is very rare that the board passes. The reasons for this can be quite subtle and usually stem from one or more of the following:

(i) errors in the circuit description, e.g., wrong device-type listed—SN7402 instead of SN7401 (one device contains NAND gates, the other NOR gates but pinouts are the same);
(ii) discrepancies between the actual circuit on the board and the circuit diagram;
(iii) the presence of real faults on the board;
(iv) generation of "illegal" input sequences in the test program*;

---

*An example of this occurs with the SN7474 D-type flip-flop. This device has both preset and clear inputs. If both these inputs are set low at the same time, the Q and $\bar{Q}$ outputs will both be held high. If preset and clear are then returned high simultaneously (or within the minimum timing period of the simulator), the resulting output status of Q and $\bar{Q}$ is indeterminate. The programmer may have no knowledge of the situation if the flip-flop is buried in the circuit.

(v) wiring errors in the board-tester interface;
(vi) inaccurate simulator models.

4.4 Types of Fault Simulator

In the previous section we discussed the overall features of a simulator-based system to support test-program generation and evaluation. At the heart of the system lies the simulator and various methods are used to increase efficiency and hence decrease costs. These methods are discussed in this section.

4.4.1 Serial Fault Simulator

This is the earliest and simplest form of fault simulator. The circuit is modelled as a normal fault-free circuit but with a single logical fault inserted. This fault overrides the fault-free behavior.

This type of simulator requires one complete pass through the simulation per fault insertion and is inefficient compared to the other three forms of simulator.

4.4.2 Parallel Fault Simulator

This is the most common type of simulator. The principle is to simulate n copies of the network simultaneously, where 1 copy represents the fault-free behavior and the other $(n-1)$ copies represent $(n-1)$ different faulty versions. The value of n is usually governed by the wordlength of the host computer.

To illustrate the principle of operation of a parallel fault simulator, consider again Example 3 (Figure 4.8) for the test input conditions $x_1 = x_2 = 1$, $x_3 = 0$.

Logical operations between various computer words are used to mirror the logical operations specified within the circuit. Each bit position in a computer word (vector) identifies with a particular version of the circuit. Position 0 corresponds to the fault-free version, position 1 to the c_1 s-a-0 version, etc. Note that since we are evaluating the fault

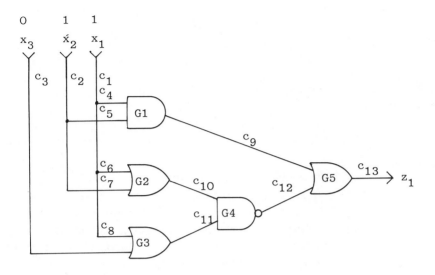

FIGURE 4.8 Example 3

cover of test input $x_1 = x_2 = 1, x_3 = 0$, there is no point in assigning a bit position to x_1 s-a-1, x_2 s-a-1, or x_3 s-a-0. For other nodes, however, we must allocate a bit position for both s-a-0 and s-a-1 as we have no knowledge of the fault-free value yet.

Assuming a wordlength of 16-bits, we will assign the following versions of the circuit to a particular bit position. Note that positions 14 and 15 are not required for this particular evaluation and that for simplicity, individual faults on the fanout branches, $c_4 - c_8$, are not considered.

Bit Position	Fault
0	Fault-free
1	$c_1/0$
2	$c_2/0$
3	$c_3/1$
4	$c_9/0$
5	$c_9/1$
6	$c_{10}0$
7	$c_{10}/1$

8	$c_{11}/0$
9	$c_{11}/1$
10	$c_{12}/0$
11	$c_{12}/1$
12	$c_{13}/0$
13	$c_{13}/1$
14	—
15	—

Figure 4.9 shows the results of the simulation.

The initial step is to create the three vectors that correspond to the values of x_1, x_2, and x_3 for each of the 14 versions of the circuit. Those values that are underlined are set into the vectors as initial conditions and represent the source fault. For example, the value in bit position 1 for the x_1 vector is 0, corresponding to the c_1 s-a-0 fault.

The procedure now is to continue creating new vectors for each of the internal nodes in the circuit. The first vector formed corresponds to node c_9 and is computed on a bit-by-bit basis from the x_1, x_2 vectors according to the logical relationship between c_9 and x_1, x_2, i.e., $c_9 =$

16-bit computer words : bit position

	0	1	2	3	4	5	6	7	8	9	10	11	12	13	14	15
$x_1 = 1$	1	0	1	1	1	1	1	1	1	1	1	1	1	1	-	-
$x_2 = 1$	1	1	0	1	1	1	1	1	1	1	1	1	1	1	-	-
$x_3 = 0$	0	0	0	1	0	0	0	0	0	0	0	0	0	0	-	-
$c_9 = x_1 x_2$	1	0	0	1	0	1	1	1	1	1	1	1	1	1	-	-
$c_{10} = x_1 + x_2$	1	1	1	1	1	1	0	1	1	1	1	1	1	1	-	-
$c_{11} = x_1 + x_3$	1	0	1	1	1	1	1	1	0	1	1	1	1	1	-	-
$c_{12} = \bar{c}_{10} + \bar{c}_{11}$	0	1	0	0	0	0	1	0	1	0	0	1	0	0	-	-
$c_{13} = c_9 + c_{12}$	1	1	0	1	0	1	1	1	1	1	1	1	0	1	-	-

Compiled vectors (inserted faults underlined)

FIGURE 4.9 Results of Parallel Simulation

$x_1 x_2$. As before, the appropriate fault value overrides the calculated value ($c_9/0$ in position 4, $c_9/1$ in position 5).

Continuing in this way, the final vector for the value of c_{13} is derived. The fault-cover of the test can now be identified as those faults corresponding to the bit positions whose observed value is opposite to the fault-free reference value, i.e., the faults corresponding to bit positions, 2, 4, and 12*.

The exercise is now repeated for the next set of test input values.

4.4.3 Deductive Fault Simulator

The deductive fault simulator is more recent than the parallel simulator and works on the principle that only the fault-free behavior is simulated but that all faults that are so far detectable on an internal or terminal node are deduced simultaneously and carried along at each stage. The mechanism is illustrated by Figure 4.10.

Each element in the circuit carries with it the normal fault-free value on its inputs (called the true value) plus a list of faults that would change the fault-free value if they were present. This list of faults comprises earlier faults transmitted up to the node in question plus the fault generated at the node (equal to the node stuck-at the complement of its true value).

In Figure 4.10 the true value on input c_{10} is 0 and the occurrence of any of the faults number 1, 2, 3, or 6 would change this value from 0 to 1. Similar interpretations can be made about the fault lists associated with the other three inputs c_{11}, c_{12}, and c_{13}.

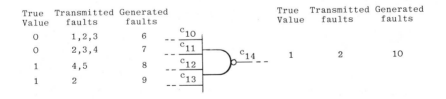

FIGURE 4.10 Deductive Simulation

*This particular test input corresponds to the conditions discussed earlier in Figure 2.7 (Chapter 2).

The deductive simulator now computes the normal fault-free value on the output c_{14} to be the true value of 1 and also looks at the input faults to see which, if they occurred, would modify the true value of the output. For this example, the only way the output will change is if the values on inputs c_{10} and c_{11} both change from their true value of 0 to a fault value of 1 *and* the true values on c_{12} and c_{13} are unaffected.

If we define the fault-lists on lines $c_{10}, c_{11}, c_{12}, c_{13}$, and c_{14} to be the sets F10, F11, F12, F13, and F14 respectively, then these conditions can be expressed quite neatly using set theoretic operators, viz:

$$F14 = (F10 \cap F11) \backslash (F12 \cup F13)$$

where $\cap$ denotes the intersection between two sets

$\cup$ denotes the union of two sets

$\backslash$ denotes the relative complement, i.e., X\Y defines the set whose members are those elements present in X but not present in Y

i.e., for the example above:

$$F10 \cap F11 = \{2,3\}$$
$$F12 \cup F13 = \{2,4,5,8,9\}$$

and therefore F14 = $\{3\}$

The result in this case is that fault number 3 is the only fault that satisfies the conditions for forward propagation; fault number 1 only changes c_{10}; fault number 4 only affects c_{11} (and c_{12}). Fault number 3 therefore is transmitted to the output line, a new generated-fault number 10 (c_{14} s-a-0) added, and so the process continues.

Figure 4.11 demonstrates the complete process for the example used earlier to demonstrate parallel simulation (Fig. 4.8). As before, the test input is $x_1 = x_2 = 1, x_3 = 0$ and the fault numbers correspond to those listed earlier.

4.4.4 Concurrent simulation

Concurrent simulation is a variation of the deductive simulation technique. The variation is in the method of implementation rather than with the principle and has to do with the manner in which fault-list data

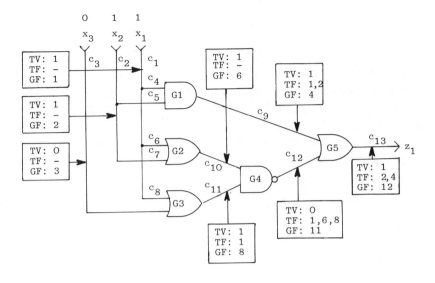

TV: True Value
TF: Transmitted fault
GF: Generated fault

Final Fault-cover = Faults 2,4,12 as before.

FIGURE 4.11 Result of Deductive Simulation

are updated as successive test input vectors are evaluated. In the deductive technique, each input vector is considered a separate entity. Although fault lists are passed on, any change in activity caused by a change in the input vector values usually means a complete reassessment of each of the nodal fault lists irrespective of whether the update has caused the data to change. On the other hand the concurrent simulator only updates lists if they are different. In other words, updates are restricted to nodes that change.

Theoretically therefore, concurrent simulation should be faster than deductive simulation but, in practice, the increase in speed is dependent on the type of circuit (feedback pattern particularly) and on the amount of new activity generated by each vector change. Concurrent simulation requires more storage than deductive simulation for the fault-lists associated with each node but is currently considered the best method for evaluating the fault-cover of circuits containing devices whose behavior is defined by a high-level functional description.

4.5 Final Comments on Simulation

Simulator-based systems of the type described in this chapter form the mainstay of test-programming in industry. There is no doubt that they have a valuable role to play, but they should be easy to use, inexpensive, and accurate in their results. Also, the interactive facilities should present helpful information in an easy-to-read format. In the end, there is always a trade-off between the level of sophistication of the facilities and the running costs.

Chapter 5
Test Application and
Fault-Finding

Chapter 5 considers the third major activity of testing—application of the test followed by fault diagnosis in the event of board failure. Essentially there are two ways of locating the cause of failure. The first is to analyze the information available at the driver/sensor pins alone—this is called "pin-state" testing—and the second is to supplement this information by further observation of internal nodal values via a hand-held or machine-driven guided probe.

5.1 Pin-State Testing

Pin-state testing, as the name suggests, means that the state of each primary input and primary output interfacing with a driver/sensor pin is checked by the tester as the test proceeds. The test program specifies the expected state of each input/output pin and the driver/sensor electronics interprets this specification and checks against what is happening on the pins themselves. An inconsistency, if it occurs, is deemed to be caused by an error condition on the board; an appropriate fault message is then printed out on a line printer or visual display terminal.

   The problem, of course, is that the amount of information about the possible cause of the fault is very limited. All that is known is the line number at which the fault was detected and which outputs on the board were found to be wrong. Hence the message "Fault at line *xyz,*" followed by the symbolic signal names of the incorrect outputs. In most cases, however, the fault is not on the output itself but is located elsewhere within the circuit. What the test has done is to set up internal logic values such that, if a particular fault is present, the effect of that fault is propagated to one or more of the tester-observable points, that is, to the primary outputs connected to the driver/sensor pins. The question, however, remains: what is the actual cause of the failure? One solution is for the test programmer to build in programmed messages that print out possible causes of failure; the other is to make use of the fault dictionary produced earlier by the logic simulator.

Consider first the programmed message technique. This is obviously limited by:

(i) the test programmer's understanding of what and where the fault might be, and
(ii) the fundamental limitation of equivalent faults.

The technique does have some value for certain categories of fault, however, and most test-programming languages contain appropriate "test-and-branch" instructions, e.g., a GOTO IF FAULT statement, which branches to a program PRINT statement if a certain test response is not observed.

5.2 Use of a Fault Dictionary

A more satisfactory solution to the diagnosis problem is to make use of a fault dictionary created by a logic simulator. Essentially, the operating system of the tester uses the fault dictionary as a look-up table. Depending on the organization of the dictionary and on the strategy employed by the tester, the diagnosis can be quite accurate (subject to the limitation of equivalent faults), and will be illustrated with the circuit shown in Figure 5.1.

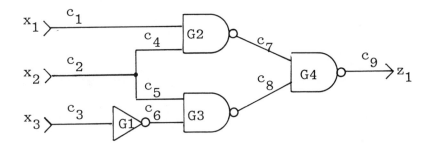

FIGURE 5.1 Example 11

This circuit is fully tested for any single stuck-at-0 and stuck-at-1 failure by the four tests T2, T3, T4, and T7 where:

	x_1	x_2	x_3	z_1
T2 =	0	1	0	1
T3 =	0	1	1	0
T4 =	1	0	0	0
T7 =	1	1	1	1

The corresponding fault-cover for each test is contained in the following fault dictionary:

Test	Fault-free Response	Fault cover
T2	1	$c_2/0$, $c_3/1$, $c_5/0$, $c_6/0$, $c_8/1$, $c_9/0$
T3	0	$c_1/1$, $c_3/0$, $c_6/1$, $c_7/0$, $c_8/0$, $c_9/1$
T4	0	$c_3/1$, $c_4/1$, $c_5/1$, $c_7/0$, $c_8/0$, $c_9/1$
T7	1	$c_1/0$, $c_2/0$, $c_4/0$, $c_7/1$, $c_9/0$

Note that there is some overlap in the individual fault-cover entries. For example, $c_9/0$ is detected both by T2 and by T7. If the circuit passes T2 but subsequently fails T7, then $c_9/0$ at least can be eliminated as a possible cause.

Figure 5.2 shows the full decision tree that might be generated,

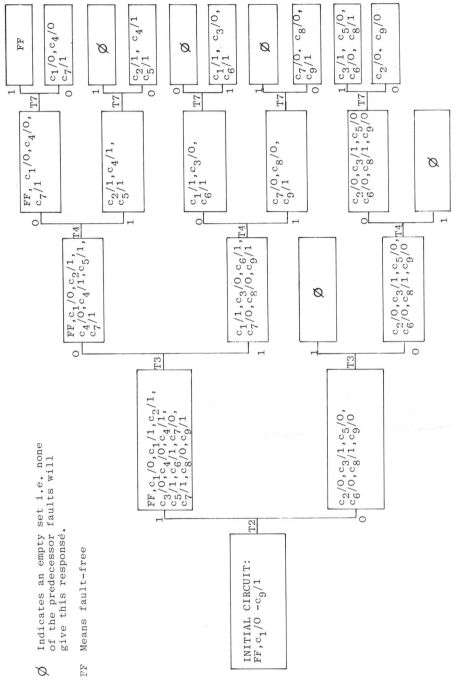

FIGURE 5.2 Fault-Location Decision Tree

104

on-line or a priori, from the information contained in the fault dictionary, assuming the order of test application to be T2, T3, T4, and T7. Note the ambiguity in all the diagnoses if the circuit is not fault-free (denoted by FF). Some of these ambiguities are caused by the fundamental limitation of equivalent faults around each of the four gates whereas others are not so obviously equivalent. Exercise 5.1 is based on this example.

Exercise 5.1

Prove that none of the fault sets listed in Figure 5.2 after the application of the last test, T7, can be split further by the application of any of the as yet unused tests—T0, T1, T5, or T6.

Exercise 5.1 : Outline Solution

The simplest (but not the only) way of proving this result is to calculate the truth table for each version of the circuit and hence show that certain sets of faults exhibit identical effect on the value of the primary output for all possible test-input combinations. This set of truth tables is shown in Figure 5.3 and, by inspection, each separate set of faults shown in Figure 5.2 can be seen to be equivalent.

---

x_1	x_2	x_3	FF	$c_1/0$	$c_1/1$	$c_2/0$	$c_2/1$	$c_3/0$	$c_3/1$	$c_4/0$	$c_4/1$	$c_5/0$	$c_5/1$	$c_6/0$	$c_6/1$	$c_7/0$	$c_7/1$	$c_8/0$	$c_8/1$	$c_9/0$	$c_9/1$
0	0	0	0	0	0	0	1	0	0	0	0	0	1	0	0	0	1	0	1	0	1
0	0	1	0	0	0	0	0	0	0	0	0	0	0	0	0	1	0	1	0	0	1
0	1	0	1	1	1	0	1	1	0	1	1	0	1	0	1	1	1	1	0	0	1
0	1	1	0	0	1	0	0	1	0	0	0	0	0	0	0	1	1	0	1	0	1
1	0	0	0	0	0	0	1	0	0	0	1	0	1	0	0	1	0	1	0	0	1
1	0	1	0	0	0	0	1	0	0	0	1	0	0	0	0	1	0	1	0	0	1
1	1	0	1	1	1	0	1	1	1	1	1	1	1	1	1	1	1	1	1	0	1
1	1	1	1	0	1	0	1	1	1	0	1	1	1	1	1	1	0	1	1	0	1

OUTPUT VALUES FOR SINGLE FAULT CONDITIONS

FIGURE 5.3 Truth-Tables for Example 11

Incidentally, this example also demonstrates the fact that fault collapsing, as described in Chapter 2, is not guaranteed to find all sets of equivalent faults. In the example above, an application of the fault-collapsing procedure would not have produced the three sets:

$$\{c_2/1,\ c_4/1,\ c_5/1\}$$
$$\{c_1/1,\ c_3/0,\ c_6/1\}$$
$$\{c_2/0,\ c_9/0\}$$

Of these, the first and third can be attributed to the fanout/reconvergent properties of the circuit, for which there are no hard and fast rules. The middle set, however, is more devious. The relationship $\{c_3/0,\ c_6/1\}$ can be attributed to the 2-way property of the inverter but the reason for the further addition of $c_1/1$ is not so obvious. It is, in fact, an example of ''functional'' fault equivalence. What this means is that no matter how this function is implemented, a stuck-at-1 on input x_1 will always be equivalent to a stuck-at-0 on input x_3. This is proved by observing the effect of each fault on the function, as expressed by its Boolean equation.

Fault-free functions:

$$z_1 = x_1 x_2 + x_2 \bar{x}_3$$

For x_1 s-a-1, i.e., $x_1 = 1$:

$z_1(x_1/1) = 1 \cdot x_2 + x_2 \bar{x}_3$

$\qquad = x_2$

For x_3 s-a-0, i.e., $x_3 = 0, \bar{x}_3 = 1$:

$z_1(x_3/0) = x_1 x_2 + x_2 \cdot 1$

$\qquad = x_2$, i.e., same as $z_1\ (x_1/1)$ above

Irrespective of how the function is implemented therefore, this functional equivalence between the two faults will always be true.

In general, it is not practical to search for fault equivalences of this type in the fault-collapsing exercise. To do so involves derivation of the Boolean function followed by observation of the effect of inserting various faults as above. The amount of work involved does not normally justify the end result and it is difficult to extend the concept to circuits containing stored-state devices and feedback.

A final comment about the use of fault dictionaries. In the example just discussed, it was assumed that all four tests were applied irrespective of the outcome of each test. Thus, it was possible to refine the degree of accuracy of the diagnosis.

An alternative strategy is to print out a diagnosis at the first point of failure. For example, if the circuit passes the first test, T2, and fails the next test, the diagnosis would be to one of six faults. For a long sequence of tests such a strategy has some advantage but incurs a penalty of loss of diagnostic resolution. As usual, the compromise is between speed of diagnosis and accuracy.

In summary therefore, attempting a diagnosis purely on the values observed at the primary output pins, with or without the benefit of a fault dictionary, can lead to a reasonably accurate diagnosis. It will always be limited, however, by the fault-equivalence relationships that exist in the circuit, and some method for overcoming this limitation is desirable. Unfortunately, this can only be achieved by further observation of the values on internal nodes in the circuit as the sequence of test progresses. This observation is made via a guided probe placed according to programmed or automatic instruction; known variously as "guided probe" or "signature" testing. The following section presents the principle underlying this method of diagnosis.

5.3 Principle of Signature Testing

To explain the principles of signature testing, use will be made of an earlier example. The circuit and timing diagrams for the set of input/output test patterns are shown in Figures 5.4 and 5.5.

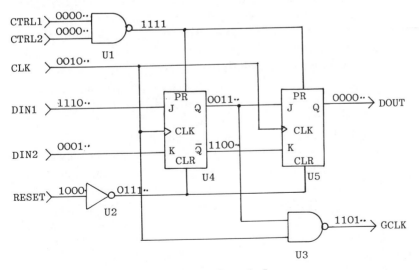

FIGURE 5.4 Example 5

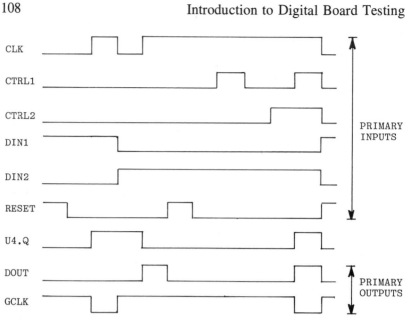

FIGURE 5.5 Timing Diagram (Example 5)

Each logic-carrying connection (node) now has a binary string of 0's and 1's associated with it. These strings contain the logic values expected on that node as the sequence of test patterns progresses; the patterns have been derived from the timing diagram. A knowledge of the binary strings together with observation via a guided probe allows identification of the faulty nodes. Furthermore, as the probe traces back through the circuit from one of the primary outputs at which the fault was detected, it becomes possible to identify the actual cause of the fault. An example of this is shown in Figure 5.6.

In this diagram, an open-circuit fault exists on the J input to the second J-K flip-flop, device U5. Because these are TTL devices, this fault will cause the J input to be permanently high, that is, stuck-at-1, and this will modify the behavior of the U5 flip-flop. When the test program is run, therefore, the tester will observe a discrepancy between the expected fault-free value on DOUT and the observed value. The question now is whether the fault really is on DOUT or whether it lies further back in the circuit.

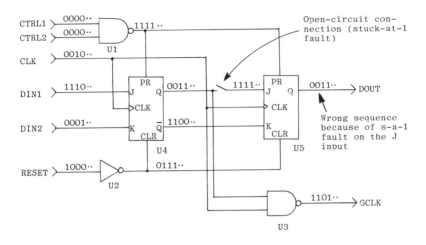

FIGURE 5.6 Fault Location by Guided Probe

To answer this question, the program is recycled, with the guided probe now used to observe the sequence of binary values on each input to the U5 flip-flop, checking what is observed with what should be observed. In this case, the guided probe would observe correct values on all inputs to U5 except the J-input. This indicates that the fault is not on DOUT but rather that it is further back in the circuit. The next stage is to use the probe again to check the sequence of binary values on the device output that connects directly to the J-input of U5, that is, on the Q-output of U4. In this case, the observed sequence compares correctly and the diagnosis is to the connection between the Q-output of U4 and the J input of U5.

In summary, the guided-probe procedure is as follows:

Question	*Answer*
1. What primary output was incorrect?	DOUT
2. What device has DOUT as an output?	U5
3. Probe U5 inputs, one at a time, and compare with stored reference data:	
U5, PR input. Match?	Yes
U5, J input. Match?	No
4. What device has an output that connects directly with U5, J input?	U4, Q output

5. Probe U4, Q output
        Match?   Yes

Diagnosis: there is an open-circuit between U4, Q output and U5, J input.

    This is the principle of signature testing. The sequence of binary values for each node is called the "signature" of that node and the technique is essentially based on the algorithm "find the device whose output-node signature is incorrect but whose input-node signatures are all correct." (In this context, a connection from a source point to a destination point can also be considered a device.)

    There are, therefore, two new and essential requirements for signature testing. The first is a guided-probe database to direct the positioning of the probe, the second is a predefined set of all nodal fault-free signatures to provide reference data on a look-up basis.

    The database is a topological description of the board, i.e., a description of the devices, identified by their physical location or symbolic identifier names, and of the interconnections between them. Usually, a topological description already exists, having been created for the logic simulator, i.e., the image. If the description does not exist, then the information needed by the algorithm is:

    (i)   the device identifier;
    (ii) the device inputs;
    (iii) the device outputs;
    (iv) the connections that exist between devices.

    The other requirement is a knowledge of the fault-free signature on each node corresponding to the predefined sequence of test-input patterns. These signatures can either be learned from a known-good-board or deduced from the simulation results. In practice, both methods are used but the preference is toward learning direct from a known-good-board. This is an activity that is carried out either by the test programmer or by the support staff. (Chapter 7 deals further with this activity.)

    For the purpose of our general discussion on signature testing, it will be assumed that nodal signatures can be deduced fairly readily and stored away (usually with the image) ready for use as a basis for the guided-probe algorithm. Essentially therefore, the data exists to allow an answer to be found for each question listed by the probing algorithm. We turn now to a more practical aspect of signature testing—the format of the signatures themselves.

5.4 TC and CRC Signatures

There is a serious practical limitation to the use of binary string signatures as shown in Figure 5.6. The length of each binary string is determined by the number of tests in the program and, for any real board, could become excessively long, thereby requiring a considerable amount of storage.

The solution to this problem is not to use the binary string itself, but a compressed version of it. There are several ways of compressing this data. One is to store the number of times each node changes its value from either 0 to 1 or 1 to 0 as the test proceeds from start to finish. This is called a "transition-count" (TC) signature and has certain attractions from the tester point-of-view. The tester simply counts the number of changes on the observed node and compares this number with the stored correct value.

One objection to TC signatures is that it is possible for a circuit to contain a fault which alters the binary sequence but not the transition count. An example of this is shown in Figure 5.7.

Figure 5.7(a) shows the 5-bit binary strings corresponding to both the fault-free response and to the $c_1/1$ response. The corresponding TC signatures are shown in Figure 5.7(b). The signature on z_1 does not change despite the fact that the output string has changed (from 00011 to 00111). The fault would not be detected therefore.

An alternative and more popular signature type is the "cyclic-redundancy-check" or CRC signature. The principle of this technique is to feed the binary string data into a feedback shift register whose final content is determined by both the nodal input data and by what has been fed back. An example is shown in Figure 5.8.

Consider the following sequence of binary data sampled from a node on the board:

$$0 \quad 1 \quad 0 \quad 0 \quad 0 \quad 1 \quad 1 \quad 0 \quad 0 \quad 1$$
$$\uparrow \qquad\qquad\qquad\qquad\qquad\qquad \uparrow$$
First sample   Last sample

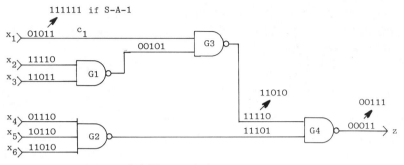

(a) Binary strings

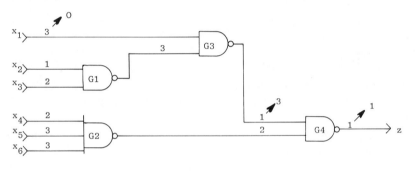

(b) Transition counts

FIG. 5.7   TRANSITION COUNT SIGNATURES

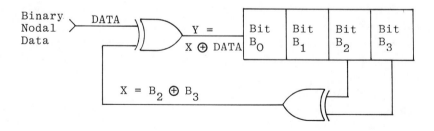

FIGURE 5.8 Feedback Shift-Register

Each time a new sample is taken, the information in the shift register is moved one place right.

The sequence of values stored in the register as each sample is made is shown in Figure 5.9.

The final signature of 0111 is the CRC signature for the node that generated the binary sequence above.

In practice, the feedback shift register is usually 16-bits long, thereby allowing up to 65,536 different signatures. What matters of course, is not so much the absolute number of different signatures, but whether a fault condition on a node will actually cause a different signature to be generated. The nature of the feedback around the shift register makes this highly probable and, in this respect, CRC signatures are significantly better than TC signatures. The figures are as follows. A 16-bit CRC signature is guaranteed to change from its fault-free value to some other value for any single-bit deviation in the input data-stream. The general probability of a sequence of multiple-bit deviations inducing an incorrect signature quickly approaches an asymptote of 1 in 2^{16} (0.99998) as the length of the bit stream exceeds the length of the feedback shift register.

By comparison, TC signatures cannot be guaranteed to catch every single-bit error and, indeed, miss $(m-1)/2m$ such errors where m is the length of the sequence. This value tends to 50% as m gets large.

Nodal data input (DATA)	Value on X $= B_2 \oplus B_3$	Value on Y $= X \oplus DATA$	Value held by B_0	B_1	B_2	B_3
S-reg initialised	0	0	0	0	0	0
0	0	0	0	0	0	0
1	0	1	1	0	0	0
0	0	0	0	1	0	0
0	0	0	0	0	1	0
0	1	1	1	0	0	1
1	1	0	0	1	0	0
1	0	1	1	0	1	0
0	1	1	1	1	0	1
0	1	1	1	1	1	0
1	1	0	0	1	1	1

FIGURE 5.9 Shift-Register Sequence

Also, the asymptotic performance of TC signatures for multiple faults is generally not so favorable as the figure quoted above for CRC signatures.

For these reasons therefore, CRC signatures are more popular than TC signatures.

5.5 The Loop-Breaking Problem

The guided-probe signature-testing technique for locating a fault works very well for acyclic circuits, that is, for circuits that do not possess structural feedback. When feedback exists in the circuit, creating a closed loop, it is possible for the effect of a fault to travel round the loop, eventually corrupting all the signatures in it. The problem then is to identify which node is causing the corruption. This, a classic fault location problem, is called the "loop-breaking" problem. It is illustrated by Figure 5.10 for part of the circuit discussed in Chapter 3 (Example 8).

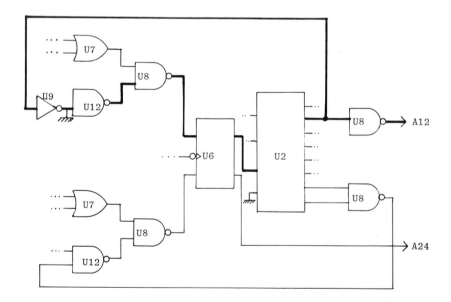

FIGURE 5.10 Loop-Breaking Problem

The fault-location algorithm is:

"Find the device with an incorrect output signature but whose input signatures are correct."

In this particular case, a s-a-0 fault on the lower input to gate U12 will cause all signatures around the loop, shown in heavy line, to be incorrect. The question is: starting from primary output A12, how can the probing algorithm determine which is the cause and which is the effect?

Some proprietary solutions to this problem exist but even these are not guaranteed to produce correct diagnosis every time. In general therefore, the problem still exists as a theoretical problem but tends to be solved practically by some mechanism for preventing fault-effect propagation completely round the loop. Techniques to achieve this include:

(i) actual blocking of fault-effect propagation by means of inhibit levels on gates (not possible in Fig. 5.10 if the effect is to be observed at A12, but possible if detection takes place on A24);

(ii) use of additional "flying" leads from unused tester sensor pins to strategic nodes on the board;

(iii) provision of on-board facilities to open-circuit feedback paths for testing and fault-location purposes. (This is discussed further in Chapter 9 on guidelines for testability.)

Chapter 6
Functional Board Testers

Now that we have some awareness of the detailed problems of generating, evaluating and applying tests, we can look more critically at the tester itself and the automatic test-system environment within which it resides. This chapter considers first the different types of tester that exist to support the various test requirements of digital boards and their components and then concentrates on the functional board tester.

6.1 Digital Equipment Test Requirements

There are three distinct phases in the life-cycle of a digital system: design, manufacture, and field-use. Each of these phases requires some form of testing, with or without repair, as illustrated in Figure 6.1.

During the design phase, the prime concern is to validate the logical design and its implementation. Emphasis is on design validation rather than on proving or disproving the existence of a fault.

This emphasis changes once the design reaches the manufacturing production line. The emphasis is then on testing for faults. For a modern digital system based on an assembly of printed-circuit-boards, a variety of test options present themselves as follows. Each type of tester is known generically as an Automatic Test Equipment (ATE). The complete test system (including software support, management system, etc.) is known as an Automatic Test System (ATS).

117

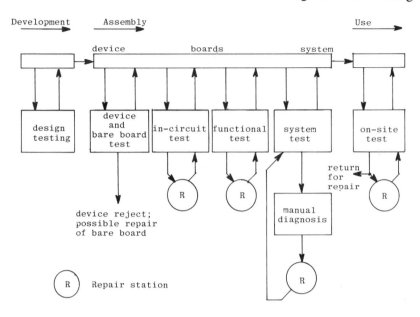

FIGURE 6.1 Board Test Requirements

6.1.1 Device Testing

Tests here can range from simple functional checkout tests, e.g., truth-table/transition-table checkout for SSI and MSI devices, to full dc and ac parametric tests with temperature cycling and supply-line tolerancing. Figures 6.2–6.5 show a selection of modern device testers. Some are general-purpose testers (Figs. 6.2 and 6.3), whereas others (Figs. 6.4 and 6.5) are designed specifically for memory devices.

6.1.2 Bare-Board Testing

The objective is to check the connectivity of the on-board copper track before components are mounted onto the board. (See Figure 6.6.)

A common technique for doing this is to make use of a "bed-of-nails" test-head fixture which, as the name suggests, is a multi-pin probe system that makes contact over the board and which is pro-

The Tektronix S-3270 is a computer-controller ATS designed for testing and verifying the performance of a wide variety of modern integrated circuits (digital, analogue and hybrids such as ADCs, DACs). It is capable of applying functional tests, dc parametric tests, ac parametric tests and linear tests at speeds up to 20MHz.

FIGURE 6.2 Tektronix S-3270 Device Tester (Courtesy Tektronix)

The Fairchild Sentry Series 20 is a general-purpose LSI test system, capable of testing at up to 40 MHz. Functional tests are performed at the programmed rate by inputting data to the D-U-T and comparing its outputs with expected values. Test patterns are held in a local memory in the high-speed controller.

FIGURE 6.3 Fairchild Sentry Series 20 Device Tester (Courtesy Fairchild)

grammed to test for various short-circuit and open-circuit connectivity patterns. Often, the tester learns these patterns automatically from a known-good-board and is then able to test other boards against this standard. A discrepancy is diagnosed as either open-circuit or short-circuit.

6.1.3 In-circuit Testing

This form of tester seeks to test individual components on the board, again via a programmed bed-of-nails fixture. (See Figure 6.7.)

The Accutest System 7800 is a 50 MHz test system designed for very high speed testing of memory devices. The system can be supplied with two test heads, each capable of testing static and dynamic RAMs up to 64K x 16 bits and 1M x 4 bits, with or without multiplexed address and data pins. ROMs and EPROMs can also be tested.

FIGURE 6.4 Accutest System 7800 Memory Tester (Courtesy Accutest)

In particular, the value of a discrete component such as a resistor can be measured. Alternatively, a logical gate can be exercised and checked against its truth table.

The main problem of course is that the component being measured must be electrically isolated from other components on the board. This is carried out by a technique called "guarding." (See Figure 6.8.)

In this diagram, Z_x is the unknown impedance (component-under-test) and Z_1, Z_2 the unwanted parallel impedance paths. By connecting both these paths to the guard point G, which is at ground potential, virtually all the current flowing into the operational amplifier comes from the V_s source and passes through the component-under-test. (The loading effect of Z_1 is negligible because of the low source impedance of the power supply V_s. Also, Z_2 is earthed at one end and is at virtual earth at the other.)

The Macrodata M-1 memory tester is capable of testing static MOS, bipolar or ECL RAMs up to 64K x 8 bits at speeds of 25 MHz. Dynamic MOS RAMs can be exercised at 20 MHz. The system is based on the DEC LSI-11 controller.

FIGURE 6.5 Macrodata M-1 Memory Tester (Courtesy Macrodata)

The Teradyne N151 is a tester for continuity of bare-boards. The bed-of-nails fixture is
shown to the left of the photograph. Data from the probes is interfaced to the tester
(centre) by a multiplexed "daisy-chain" interface.

FIGURE 6.6 Teradyne N151 Bare-Board Tester (Courtesy Teradyne)

Under these conditions:

$$Z_x = -R_{ref} \times \frac{V_s}{V_0}$$

In practice, the situation may not be quite so simple as this diagram
suggests and "extended guarding" techniques exist to overcome errors
due to other impedances (e.g., the voltage source impedance, guard bus
impedance, lead impedance, etc.).

To test the behavior of logic devices, the board, or at least the
device-under-test (D-U-T), must be powered up. Testing is then carried
out by using the bed-of-nail probes to drive the D-U-T inputs through a
programmed sequence. Other probes are used to sense and compare the
response. As before, the main problem lies in isolation from other
logical devices that are also powered up and driving the D-U-T. A

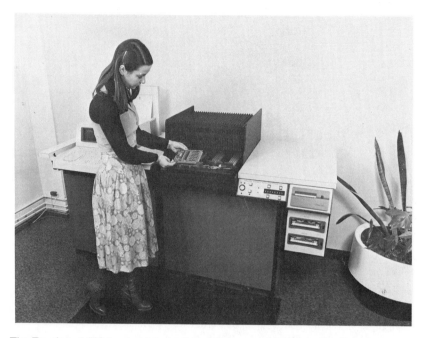

The Teradyne L529 in-circuit tester is used to pre-screen production boards prior to functional board test. Its aim is to find gross manufacturing faults. The system can be programmed by a learn technique to minimize programming time. The bed-of-nails fixture is mechanical and uses a matrix of pressure rods to maintain good contact between board and probes.

FIGURE 6.7 Teradyne L529 In-Circuit Tester (Courtesy Teradyne)

common technique for overcoming this is to stimulate the D-U-T inputs with pulses that are long enough to create input values of 0 or 1 but not long enough to cause damage to predecessor devices. Typically, such pulses are 10μ s long for TTL devices.

6.1.4 Functional Testing

Functional testing complements in-circuit testing inasmuch as in-circuit testing isolates and tests individual components and devices on the board, whereas functional testing tests the overall function of the board assembly, i.e., the collection of devices and their interconnects. That is

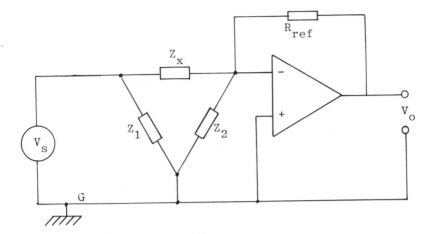

FIGURE 6.8 In-Circuit Testing

the theory at least but, as we have seen, writing the test patterns is not easy.

In general, however, a functional tester will only access the board through its normal edge-connector or test-access connector. Access to internal nodes for diagnostic purposes is normally limited, and is via a single or multi-point guided probe, as described in the previous chapter. In practice this restriction is often relaxed for reasons usually to do with accuracy of diagnosis but, in principle at least, test stimulus and observation should be from the edge connector alone.

Figures 6.9–6.11 show examples of functional testers at the top end of the range. All the testers shown are capable of testing both analog and digital boards.

6.1.5 System Testing

By the time each board has been tested to an acceptable standard and assembled to form a system, the test requirement moves back to one of design testing—this time at a system level. General-purpose automatic test systems are not made for this level of testing. Each system dictates its own system-level test requirements based on the original design specification, and a common strategy is to make use of a "substitution

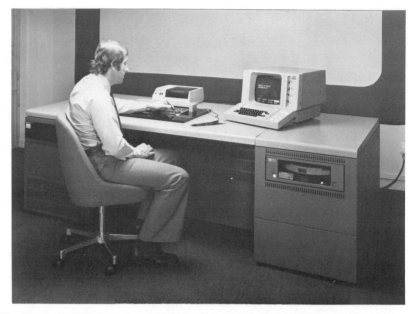

The Membrain 7730 is a range of functional test systems for digital boards but with an analogue capability. Features include programmable test patterns, stored-response and signature testing, and a selection of fixed and programmable driver/sensor pins. Fault isolation is by guided clips and probes, including a current-sensing probe called the "Flo-tracer" (shown in use in the photograph above and in more detail in Figure 8.8 in Chapter 8).

FIGURE 6.9 Membrain 7730 Functional Board Tester (Courtesy Membrain)

rig." This is a working system into which a new board, substituting for the working board, can be loaded. The new board is then tested against certain system-level acceptance tests (benchmark tests).

6.1.6 Field Testing and Servicing

Until recently, the main strategy for field servicing was to swap boards until the system fault cleared and then to send the suspect boards back to a central repair center. Such a center would probably be equipped with the same functional tester as used on the production line and the same functional test program would be used to diagnose the board so that a

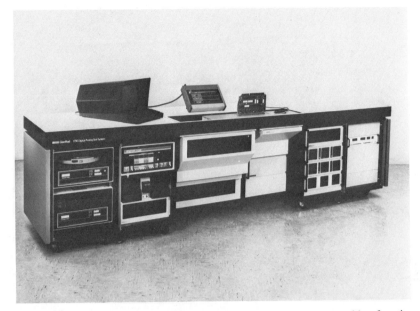

The GenRad 1796 is a combined analogue/digital functional tester capable of testing boards up to 1.5 MHz. The system is supported by a software simulator, called CAPS, and features on-line analysis of the B-U-T's response to attempt diagnosis without probing. If this fails, guided probe facilities exist which, in conjunction with the CAPS-generated fault dictionary attempts to find the fault with the minimum number of probes.

FIGURE 6.10 Genrad 1796 Functional Board Tester (Courtesy Genrad)

repair could be carried out. There are two disadvantages to this scheme. The first is the requirement for a large inventory of spare boards, the second is the fact that many supposedly defective boards are subsequently found not to be faulty. Such boards are referred to as "no-fault-found" boards.

To offset these disadvantages, board-swapping is usually the fastest method for bringing the digital system back to a working state and, in the end, this is what counts from the end-user's point of view.

Recently, however, there has been a spate of new testers which, although physically small and therefore portable, have the same capabilities for go/no go testing and almost the same fault-location capabilities as the larger production-line testers. Furthermore, if com-

The Computer Automation 4900 functional tester can test hybrid boards at speeds up to 2 MHz, and has a high speed clock capability of 10 MHz. The system is based on an LSI-2/20 16-bit processor and features twin floppy discs together with a 10M byte moving head disc drive.

FIGURE 6.11 Computer automation 4900 Functional Tester (Courtesy Computer Automation)

patibility between the production-line and field-service testers exists, the same test program can be used. An example of a portable tester of this type is shown in Figure 6.12.

The advent of these testers has allowed decentralization of the field-servicing repair requirement. In effect, the field-service engineer is now able to visit a customer site and test each board against a production line functional test until the faulty board is identified. At that point, the engineer can either carry out a repair on-site, or he can simply swap the board and send the defective board back to a repair depot as before.

There are many factors which influence a decision to centralize or decentralize field test-and-repair. Some of these are as follows:

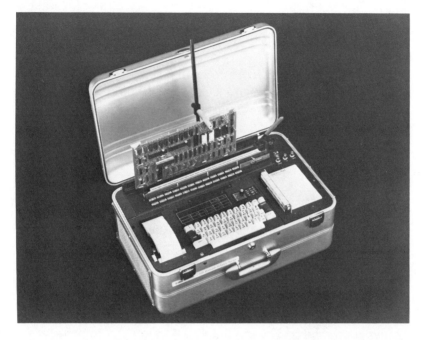

The GenRad 2225 portable service tester is a tester designed specifically for field service diagnosis of digital PCBs. The tester is based on an LSI-11 controller and contains 192 programmable driver/sensor pins. Fault diagnosis is by a software-controlled guided probe system based on CRC signatures.

FIGURE 6.12 Genrad 2225 Field-Service Tester (Courtesy Genrad)

(i) The inventory costs of spare boards and spare devices and the ease with which a board repair can be carried out in the field.

(ii) The need to distribute and maintain test programs. (This becomes particularly complicated if the boards are subject to modifications, commonly called "revision levels" or "engineering change orders.")

(iii) The cost of supporting a repair system for boards which are eventually found to be no-fault-found.

(iv) The skill level required by the field-service engineer and the question of whether such engineers can be found and trained.

(v) The importance of maintaining the integrity of a customer's installation. This is particularly important for high-risk technical applications such as on-line process control, but it is also important for non-technical systems such as banking or record-keeping in general.

(vi) The cost, availability, and compatibility of production and field testers.

(vii) Last, but not least, the decision to decentralize test-and-repair may influence the actual design and implementation of each board in the system. An example is the Signature Analysis technique. This is a procedure for implementing built-in-test onto a board in order to simplify field servicing, particularly fault location. The technique is described in Chapter 9.

The remainder of this chapter will concentrate on functional board testers. However, much of the following discussion also relates to the other forms of tester, particularly device testers and in-circuit testers.

6.2 Major Considerations of Functional Testers

In this section we will look at a number of facets of functional testers. The order does not necessarily imply any priority of importance and, although the topics have been separated for convenience of discussion, there are obvious interrelationships.

6.2.1 General Configuration : Comparative and Stored-Program

Figure 1.7 in Chapter 1 presented the general layout of a tester. Basically it consists of a computer interfacing to the board-under-test (B-U-T) via the driver/sensor (D/S) pin interface. Like any computer system, there may be many peripheral attachments such as disc and floppy-disc memory systems, visual-display units, line printers, etc. Also, there may be other free-standing or linked programming stations to allow initial coding and editing of test programs off-line from the tester. A programming station is a functional tester without the driver/sensor interface, i.e., a normal minicomputer. It can usually be upgraded to full tester status at a later date if the testing requirement grows. An example of a programming station is shown in Figure 6.13.

Of particular interest is the method by which judgment is pronounced on the status of the B-U-T. The general method discussed so far

The GenRad 2290 multi-user programming station is designed to complement GenRad's range of in-circuit testers. As an off-line programming station, it increases productivity by allowing simultaneous program generation by up to four users, as shown here. The system is based on a DEC PDP-11 computer.

FIGURE 6.13 Genrad 2290 Programming Station (Courtesy Genrad)

in this book has been to base this judgment on a stored-program of test patterns, i.e., input stimulus and expected output response. This is the most common technique and is referred to as the "stored-program" method.

An alternative method exists, based on the availability of a known-good-board (K-G-B). The method is illustrated in Figure 6.14.

The basic idea is that the tester stimulates both the B-U-T and the K-G-B simultaneously with the same input patterns and that the K-G-B provides the reference output data against which the response of the B-U-T is compared—shown figuratively by the exclusive-OR gate in Figure 6.14. Any discrepancy in the two responses is considered to be a failure on the part of the B-U-T. Diagnosis of the fault is then carried out by a "dual guided probe" technique, similar to the single probe system. Effectively, the operator probes the same point on each board until the

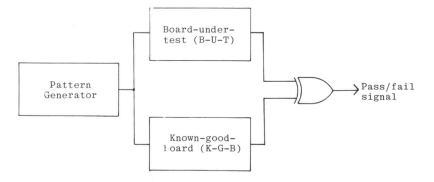

FIGURE 6.14 Comparative Testing

source of the discrepancy is found. An example of a functional board tester that employs this technique is shown in Figure 6.15.

Such a tester is referred to as a ''comparative'' tester and, in its simplest form, consists of a minicomputer (or even a microcomputer) programmed to produce a sequence of input vectors generated on a pseudo-random (P-R) basis. (Note that both boards must be initialized to the same starting condition before the pseudo-random input stimulus is started.) The procedure is to apply a P-R sequence that is long enough to create fault-propagation conditions for all the anticipated fault modes of the board. In practice there is no guarantee that this condition is satisfied and more sophisticated forms of comparative testers include simulation back-up facilities to evaluate the fault cover. Alternatively, two K-G-B's are loaded and the fault cover verified by physical fault insertion on one of them.

Comparative testers also contain a variety of P-R pattern formats that can be assigned to specific B-U-T input pins according to the function of that pin. Examples are:

(i) data sources—parallel bit streams changing at a high, medium, or slow rate (relative to a basic clock-rate frequency). Changes may be true pseudo-random (simultaneous transitions possible) or pseudo-Gray (single transition only, selected on a random basis).

(ii) clock lines—parallel square-wave bit streams at high,

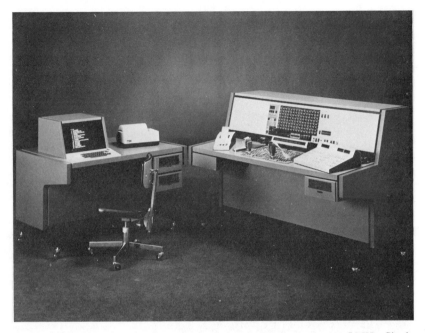

The Fluke 3040A is a comparative tester capable of testing at speeds up to 5 MHz. Single boards can be tested for go/no go using a CRC signature technique. The photograph shows the dual guided probe facility for locating a fault by comparison with a known-good-board.

FIGURE 6.15 Fluke 3040A Comparative Tester (Courtesy Fluke)

medium, or slow rates, or multi-phase, non-overlapping bit streams (for 4-phase MOS, for example).

(iii) other control lines. Bit streams that are predominantly low (or high) that can connect to other major control lines such as reset and clear inputs.

While on the subject of P-R sequences, it is worth making a few comments about the technique as a method for generating test patterns. The basic assumption is that a sufficiently high set of random but repeatable input patterns will contain a test somewhere for every fault condition on the board.

The advantages of the technique are that there is less need for detailed circuit analysis and that the P-R sequences are quickly defined.

The disadvantages are quite severe however. First, there is no guarantee that all the faults will be detected and, certainly for highly sequential circuits, it is extremely unlikely that the fault cover will reach much higher than 60%–70% on average. The remaining 40%–30% usually turn out to be the more difficult faults anyway.

The second disadvantage is that timing relationships could well be marginal, especially on boards containing monostables. This may result in one board passing the test and another failing, even though the latter is fault-free.

A third disadvantage is the need to identify the K-G-B. This is the classic "chicken and egg" problem and has to be solved by some other form of testing. Once such a board has been identified however, there is then the need to ensure that it remains a K-G-B. This may involve special storage conditions. In any event determination and maintenance of a K-G-B is an expensive process and gives rise to the alternative interpretation of the "G" in K-G-B, namely known-golden-board.

6.2.2 Driver-Sensor Pin Electronics

The driver/sensor (D/S) pins are the main interface between the tester and the B-U-T. They are therefore the most important and often the most vulnerable part of the tester. Figure 6.16 shows a basic scheme for a bidirectional D/S pin, i.e., one that can be programmed to be either a driver or a sensor, and indeed, whose driver/sensor status can be altered during the course of a test program, as would be required by a bidirectional bus, for example.

The circuit is based on the use of two operational amplifiers—one acting as the driver, the other, as the sensor. Consider first the driver operation. The 0 or 1 level required is loaded into the first rank latch by a pin-specific data load clock and when ready is transferred to the second rank and thence to the driver amplifier. This amplifier will produce one of two programmed voltage outputs, V_H and V_L, corresponding to the high and low values of the B-U-T logic devices (+5v and 0v for TTL devices, say).

When used as a sensor, the expected value is again loaded into the first rank and thence to the second rank when the data transfer clock is received. This particular clock is usually broadcast simultaneously to all

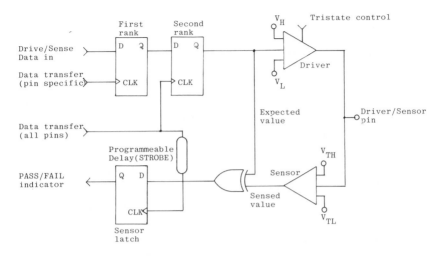

FIGURE 6.16 Driver/Sensor Electronics

pins (or maybe groups of pins), so some delay is required to allow time for the board to respond to the new driver values on other pins. This delay, called the STROBE time, is programmable and should be at least greater than the worst-case propagation delay on the board.

The sensed voltage value from the board is received by the sensor amplifier and compared with two threshold values, V_{TH} and V_{TL}. Any sensed value above the high threshold value, V_{TH}, produces an amplifier output of logic 1; any value below the low threshold value, V_{TL} produces an output of 0. Typically for TTL logic, V_{TH} is 2.4v and V_{TL} is 0.8v, i.e., the lower end of logic 1 and the higher end of logic 0.

The sensed value is compared with the expected value and, after the programmed delay, the outcome is latched into the sensor latch. A latched value of 0 indicates PASS; 1 indicates FAIL.

To avoid loading effects by the D/S system on the B-U-T when used as a sensor, it is usual to provide facilities for tri-stating the output of the driver amplifier. Effectively therefore, the board sees only the high impedance input of the sensor amplifier.

Note also that when used as a driver, the system is essentially self-checking by virtue of the driven-output being sensed and compared with the expected value.

There are many variations of detail on this scheme, one of which is to provide only one threshold reference level on the sensor amplifier.

This is called a "single-threshold" system, as opposed to the "dual-threshold" system shown in Figure 6.16. Typically for TTL, the single-threshold value would be 1.4v.

6.2.3 Test-Application Rate

The rate at which tests can be applied is obviously important for throughput reasons or for reasons connected with the desire to test a board at or near its maximum speed of operation. The main factor affecting the test-application rate is the slew rate of the D/S operational amplifiers, but the efficiency of compiled or interpreted code also plays a part. In some testers, the D/S pin electronics is controlled directly by a front-end microprocessor that interfaces between the main system computer and the D/S pins. Basically, the procedure is that the main processor acts as an overall controller but the detailed control of the tests is carried out by the front-end processor. In this way, the main processor can attend to other background duties, such as servicing user-interrupts, while the front-end processor is able to carry out its dedicated task of applying stimulus and observing response.

6.2.4 Tester/Board Interface

Ideally, we would like to be able to plug the board into the tester and commence testing. In practice, there is a need to provide a board-specific interface to cover some or all of the following requirements:

(i) to supply power and ground reference to the board;
(ii) to link special test sockets from the board to the tester;
(iii) to enable access to certain internal nodes.*

This interface, while not difficult to construct, becomes another part of the inventory required to support the testing of the board. Obviously, the more universal the interface, the easier the problems of stock and distribution.

*An example here might be the output of an on-board oscillator to which the tester must be synchronized.

6.2.5 *Test-Programming Aids*

We have already discussed the use of logic simulation as a major support for test-programming activities (Chapter 4) but, to be effective, the various aids should be available, easy to use, and helpful. This in turn requires that the tester, or programming station, should possess interactive facilities coupled with good editing facilities, and a file-management-system.

6.2.6 *Method of Fault Location*

In a repair environment, the ability to diagnose the cause of failure is of paramount importance. Incorrect or ambiguous diagnosis quickly leads to disenchantment. The method and accuracy of the diagnostic technique, therefore, should be a critical factor in deciding upon which tester to buy. Features that should be considered include:

(i) the actual method for locating the fault, e.g., guided-probe (single or multi-clip), table look-up (fault dictionary), or even a combination of both these techniques.

(ii) the speed of fault location. Procedures based solely on the use of a guided probe may take some time to reach the cause of the fault simply because of the length (number of probe points) between the source of the fault and the output at which the fault was detected. Long probing sequences (100 + probes) are not usually welcomed by repair operatives and can be very tiring, leading to error.

(iii) the ease with which probing can be carried out by relatively low-skill labor and the facilities built into the algorithm to guard against misprobed nodes and intermittent probe contact, both of which will almost certainly cause incorrect values to be observed.

(iv) the ability of the diagnostic algorithm to handle fault-effect propagation around feedback paths (Chapter 5), or even to handle multi-output devices whose outputs are dependent and unbuffered. (See Figure 6.17, for example.)

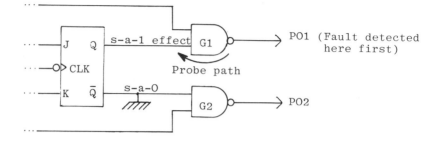

FIGURE 6.17 Incorrect Diagnosis

For this circuit the actual fault is a s-a-0 on the input to G2, but holding the $\overline{Q}$ output of the flip-flop low will cause the Q output to appear to be s-a-1. If detection is at PO1 before PO2, the probing algorithm will pronounce the Q output to be faulty, rather than $\overline{Q}$. Visual inspection of the track between Q and G1 will reveal no problem, so the most probable repair strategy will be to replace the J-K flip-flop. On retest, the board fails again with exactly the same diagnosis!

6.2.7 Flexibility

In general, testers should be able to test boards containing either TTL, CMOS, or other families of logic devices. Occasionally, a single board will contain a mix of technology. This is not usually a problem provided independent programmable power supplies exist and can be routed to the appropriate edge-connector positions. What sometimes proves difficult however, is to establish the correct driver high, driver low, and sensor threshold values for mixed-logic boards. Often these values are established for a group of D/S pins rather than each pin being independently controllable, and although this is not an insurmountable problem, it can lead to complicated set-up and interface requirements.

Also on the question of flexibility, some boards will contain a mix of analog and digital circuitry. Testing analog circuitry is a subject in its own right but basically, the tester will need facilities for generating various forms of analog signal (sinewaves, ramps, impulse, etc.) and for measuring the properties of signals received from the board (frequency, amplitude, and phase, at least).

Only digital board functional testers at the top end of the range include these types of facilities. However, many specialized analog testers exist.

6.2.8 Tester Portability and Data-Communication Facilities

With reference to the earlier comments on decentralized testing, portability may or may not be a significant consideration. If it is, however, then it is important that test programs developed for the production line environment can also be run on the portable field tester. Also, the problem of updating programs becomes significant. The simplest mechanism for updating is to issue an updated tape cassette or floppy disc. Alternatively, however, programs can be transmitted either by a hard-wired system (over an RS232C serial data line, an IEEE 488 bus, or telephone acoustic modem) or by a radio communication link. Local and remote data communication facilities may be a requirement therefore.

6.3 Test Economics

To conclude our general discussion of the use of an automatic test system, we will consider the basic economics of testing. The intention is not to present a detailed study of testing economics. That is a topic that can only be handled against a background knowledge of a specific manufacturing and field-servicing environment. Rather, the intention here is simply to comment on the more significant factors that can influence the costs and justification of testing.

6.3.1 Effect of Fault Cover on Yield

In this section we will study the effect of using a set of board tests whose ability to detect all fault conditions on the board is less than perfect. To do this we will have to make certain simplifying assumptions, but these do not detract significantly from the logic of the argument.

A common term used in digital system production is "yield." In general terms, the yield is that percentage of a collection of items that is considered to be fault-free, i.e., will work to the design specification. We will use the word in connection with components, boards, and systems.

Assume first that, on a particular board, there are n components. Components here can include not only the integrated-circuit devices but also any discrete components, the interconnections, and even solder joints. Each of these components can fail, thereby contributing to the overall failure rate of the board. For the purpose of our discussion, however, we will assume that by the time the board reaches the functional test station on the production line, the only cause of failure lies with the devices and solder joints, i.e., the bare-board test has eliminated any problems with interconnects.

Let

YD = device yield (percent fault-free)

YS = solder-joint yield

The value of YD will be determined by the degree to which devices are tested as they are bought in (goods inward test) and the confidence level that is placed on the test. Similarly, YS will be determined by the efficiency of the soldering process.

For a board containing 100 14-bit dual-in-line devices and therefore at least 1400 solder joints, the probability that each device is fault-free is YD and the probability that each solder-joint is fault-free is YS. If we assume statistical independence between any device or solder joint failure, then the probability of the board being fault-free, i.e., its yield YB, is given by:

$$YB = \prod_{1}^{n} YD \times \prod_{1}^{m} YS$$

$$= YD^{n} \times YS^{m}$$

where n = number of devices

m = number of solder joints

Assuming values of YD = 0.99 (1 device per 100 defective) and YS =0.999 (1 joint per 1000 defective), the board yield for a board containing 100 devices and 1400 solder joints is given by:

$$YB = 0.99^{100} \times 0.999^{1400}$$
$$= 0.083$$

What this figure means is that there is only an 8% probability that the board is actually fault-free, i.e., 92% probability that it is defective in some way. Assuming a uniform distribution of faults, these figures will also be true for a population of boards of this type.

Ideally, the functional test on such a board population should detect all the defective boards such that the board yield after test is 100%. In practice, the functional test may not be capable of detecting all faults, i.e., its fault cover (FC) may be less than 100% of the target set of faults.* Under these circumstances, there is a definite probability that some of the defective boards will be passed as fault-free. Denote this percentage by DB. If FC is 100% (1.000), then the percent defective boards passed is 0%. Let us assume, however, that the test program has been written to a specification that sets the minimum value of FC to 90% and has only just met this specification. (90% is a common minimum acceptable value in industry). This means that 10% of the defective boards will be passed as fault-free, i.e.

$$DB = (1 - FC)(1 - YB)$$

In the example above, and using FC = 0.90, the percent defective boards passed will be $0.100 \times 0.917 = 0.092$. In other words, of all the boards of this type that are considered to be fault-free and ready to be used in the system, 9% are in fact defective. If this figure is generally true of all the different boards in a system, then the implication is that in a system with 20 boards say, something like 2 boards will contain some fault that might eventually lead to failure, either at system test or in actual field operation. What makes matters worse is that if a suspect board is returned to the central repair depot, it may be tested with the

---

*We remarked in an earlier chapter that even a target set of all nodes s-a-1/s-a-0, supplemented by some select bridging fault possibilities, may not be sufficient to cover all the modes of failure of the board. If this is the case, the situation is even worse than that described here.

same test program as before, thereby running the risk of being returned as a no-fault-found board.

At this stage, it is useful to rework this example with changes to the yield and fault-cover figures. The exercise is detailed below.

Exercise 6.1

Making the same assumptions as in the text, calculate the improvement in the percent defective board passed figure for the following yield and fault-cover values.

Note:
$$0.99^{100} = 0.366, \; 0.999^{100} = 0.905, \; 0.999^{1400} = 0.246$$
$$0.9999^{1400} = 0.869$$

  (i) YD = 0.999, YS = 0.999, FC = 0.90
      More effort is put into the goods inward testing of the devices.
 (ii) YD = 0.999, YS = 0.9999, FC = 0.90
      More effort into both goods inward testing and the soldering process.
(iii) YD = 0.99, YS = 0.999, FC = 0.95
      Yield figures unchanged but more effort put into the fault-cover value.
 (iv) YD = 0.99, YS = 0.999, FC = 0.99
      As (iii) but even better fault-cover performance.
  (v) YD = 0.999, YS = 0.9999, FC = 0.99
      Improvement in all three factors.

Exercise 6.1 : Outline Solution

Detailed results are left for the reader but, for a general indication of how the values of YB and DB change, see Figures 6.18 and 6.19.

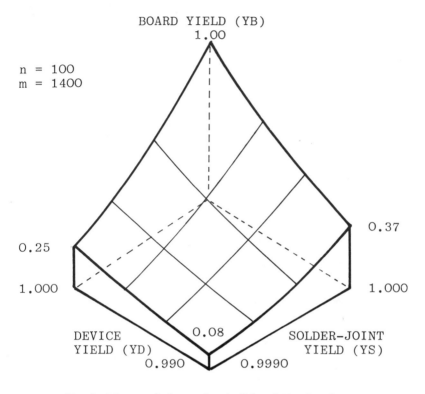

Variation of board yield with device
yield and solder-joint yield.

FIGURE 6.18 Variation of Board Yield

6.3.2 Life-Cycle Costing

One way of justifying, or otherwise, the purchase of an automatic test
system is to look at the life-cycle cost and observe its sensitivity to
various changes in the parameters. A simplified cost equation is as
follows:

$$COST = C1 + C2 + C3$$

where C1   = average cost associated with writing and evaluating test
programs for different board types

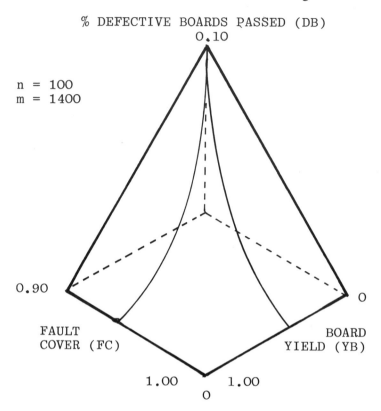

% DEFECTIVE BOARDS PASSED (DB)

0.10

n = 100
m = 1400

0.90

FAULT
COVER (FC)

BOARD
YIELD (YB)

O

1.00   1.00

O

Variation of % defective boards passed
with board yield and fault cover

FIGURE 6.19 Variation of % Defective Boards

C2 = average cost associated with the use of the test program—test application for all boards followed by fault
diagnosis on those found to be defective.

C3 = capital expenditure

In simple terms:

C1 = (P + I) B

where: P   = average programming costs per board (including the cost of any back-up facilities such as a logic simulator and the computing resources to support the simulator).

   I   = average interface costs per board. (May be significant for bed-of-nails system.)

   B   = number of different board types

   $C2 = TN + D(1-Y)N$

where T   = average test application costs per board (time and labor)
   N   = Total number of boards to be tested during the lifetime of the tester
   D   = average fault-diagnostic costs per board (time and labor)
   Y   = average board yield. $(1-Y)$ will be found defective and will require diagnosing

   $C3 = CS$

where C   = Unit cost per tester
   S   = Number of test stations

Total life-cycle cost therefore is given by:
$$Cost = (P + I)B + (T + D(1 - Y))N + CS$$

This equation, or something like it, can form the basis for cost justification of a tester for use either in a production environment or a central repair depot. In general terms, the dominant factor in a high-volume production environment tends to be the costs associated with the test-application and fault-location times, i.e., T and D. In the field-servicing environment, however, the capital expenditure term, CS, can dominate, particularly if a decentralized policy is implemented, i.e., if S, the number of test stations, becomes large.

To summarize this chapter, we can say that automatic test systems are necessary in that they provide a cost-effective tool for tackling the test-and-repair problems associated with complex digital boards. Their use should allow a figure for the production throughput of boards that is high enough to create a competitive product. There are also other spin-offs, such as continuity of support, smoother scheduling of work-flow, consistency of testing, etc.

A major requirement of ATS, however, is that it requires a management structure to manage not only the ATS, but also the many support requirements, e.g., use of simulation, device-library updates, test-program maintenance and updates, program validation, training and education, etc. There is also the problem of confidence caused by inadequate performance of the test program. A successful ATS is based on recognizing the need for a management structure and on understanding the limitations of the system.

Chapter 7
Developing and Using a
Test Program

Chapter 7 returns to the main subject of writing test programs and goes through the ten stages necessary to produce a validated and documented test program for use in a production or field-service environment. The stages are shown in outline in Figure 7.1 and are discussed further in the chapter. The discussion is presented largely from the point of view of the test programmer.

The chapter concludes with some comment on the possible problems associated with the use of the program, i.e., as seen from the user's point of view.

7.1 Stage 1 Reconstruct the Circuit Diagram

This stage is only necessary if the circuit diagram exists in fragmented form, as produced by the draughting facilities of some design automation systems. The general aims of the reconstruction are to produce a layout that suits the thought processes of the test programmer and does not disguise certain logical configurations.

An example of the problem was presented in Chapter 3 (Example 7); the discussion on that example also commented on the rules for layout. One of the most important considerations for layout is to identify

147

STAGE 1   Study schematics and develop reconstruction of circuit diagram.

STAGE 2   Develop overall test plan and prepare an estimate of the cost/time-to-completion.

          REVIEW OF ACTIVITY TO DATE

STAGE 3   Design and commission board-test interface.

STAGE 4   Code, assemble and debug the topological description of the circuit (image).

STAGE 5   Develop board-set-up and initialisation sections of the test program.

STAGE 6   Write and debug main test program.

STAGE 7   Determine the fault-cover and modify test program if target not met.

          REVIEW OF ACTIVITY TO DATE

STAGE 8   Learn and verify signatures. Modify image or test-program if necessary.

STAGE 9   Check the fault-location performance. Modify the image or test-program if necessary.

STAGE 10 Prepare release version of the program together with support documentation.

FIGURE 7.1 Ten Stages for Development of a Test Program

the true feedback paths in the circuits (from one device to another). These are the routes that will cause most of the problems both in generating fault-detection tests and in subsequent attempts to ensure correct location of the fault.

7.2 Stage 2 Develop Overall Test Plan

This stage is largely a "thinking" stage but it is quite critical to the success of the final program. Stage 2 is concerned with the preparation of the overall test plan followed, if required, by an estimate of the time-to-completion for the final test program. This estimate is used for forward scheduling of the use of the tester and for estimating the cost of writing the test program.

The following considerations affect the development of an overall test plan.

7.2.1 Test Strategy

The question of the test strategy—functional, structural, or hybrid—has been discussed in Chapter 3. In the absence of any specialized strategy dictated by the devices on the board, the hybrid functional-plus-structural approach is recommended.

7.2.2 Special Features

PCBs often contain special features that require special attention. Some examples are: onboard switches—the programmer will need to instruct the user which position to place them in; potentiometers—fully clockwise or fully counterclockwise; jumper links—which position again; on-board non-controllable oscillators—can the tester be synchronized to the oscillator or does it matter anyway; active discrete components and analog circuitry in general—how will these be tested; dynamic MOS devices with minimum as well as maximum operating frequencies—can the tester test fast enough and what about the requirement for refresh cycles; bidirectional busses—any problems with reversing the driver/sensor pins; monostables—are they directly observable and what are their periods; and so on. The list of special features is optionally large and it is really up to the test programmer to go over the circuit diagram very carefully and consider how each feature will be handled in the test program.

7.2.3 Data Sheets

If the devices are identified by a generic identifier—for example, by a series 74 TTL chip—then there should be no problem in finding the right data sheet for the device. Occasionally, however, devices will be identified by an in-house code; it then becomes necessary either to be able to relate these devices to their corresponding generic identifier or, alternatively, to have access to in-house data sheets.

7.2.4 Set-Up Tests

The next consideration is whether it is intended to include tests to establish that the board has been inserted correctly; that the power supply values are correct; and that any other special fixings have been set-up correctly. These tests, called "set-up" tests, are useful to prevent damage to the board caused by failure of the user to carry out instructions correctly.

7.2.5 Initialization

The objective of initialization is to set the board into a known start state. This means defining the logic level on all primary inputs, all primary outputs, and all internal nodes. Initialization is not simply a matter of presetting or clearing the stored-state devices on the board.

The types of problem that occur here are numerous. Example 6 in Chapter 3 illustrated one problem caused by "clever" logic design; there are many other examples. Common problems are: preset or clear lines to stored-state devices are tied internally to their enable level rather than brought out to an edge-connector position; long shift registers may need clocking to set the contents into a known state; ROM truth tables may not be available to allow knowledge of outputs for a given input; and so on.

7.2.6 Pin-State Testing or Signature Testing

Some thought should be given to the overall structure of the program from the point of view of fault resolution. Will it be wholly pin-state testing, wholly signature testing, or a mixture of both? If pin-state testing is to be used, will it require the use of a digital multimeter or of other external equipment, such as an oscilloscope?

7.2.7 Miscellaneous Considerations

The last consideration concerns a number of unrelated but important details, such as whether the programmed pin-state changes will be sequential or simultaneous—that is, whether the SKEW or BROAD-SIDE mode of testing will be used. Also, there is the question of the STROBE setting—default value or otherwise? The use of pseudo-random pattern-generation facilities may also be considered, along with other aids specific to a particular tester.

The end product is a test plan containing a flow chart of the overall test strategy together with any other comments relating to the board. It is likely that this plan will be revised as the understanding of the circuit expands but the plan gives a starting point. Also, deriving the plan forces the test programmer to think about the potential problems of testing the board. This thinking time is equally as important as writing and debugging the program and can save much effort later on. It is also helpful if the test programmer can discuss and defend the test strategy with someone else. This has two advantages: it shows an understanding of what to do and it may identify either an alternative, better way of carrying out a particular test or highlight a problem that the test programmer had not realized existed.

7.3 Stage 3 Board-Tester Interface

This is the practical stage of deciding exactly how the board will be loaded onto the tester (left or right justification in a socket, components facing operator or not, etc.) and what interface requirements are necessary. The simplest requirement is a set of leads to route power from the tester's programmable supply pins to the appropriate board edge connector positions. Other board-specific interfaces may also be required.

Whatever the requirements, these interfaces should be specified and their manufacture commissioned fairly quickly, for without them the test programmer is unable to power-up the board in order to develop and debug the program.

7.4 Stage 4 Prepare the Image

The image is the program that describes the connectivity between the devices on the board and, in that sense, producing an image should be quite straightforward. This is not quite true, however, as many problems and frustrations in the latter stages of program development can be traced to insufficient thought at the imaging stage. The image is as important as the test program and should be prepared carefully and with due regard to its prime function: to be the database for the guided probe. This means that it should be prepared so as to keep unnecessary probing to a minimum, and to make the fault diagnosis as accurate as possible. Recall that the probing algorithm is trying to find a device with an incorrect output signature but with correct input signatures. There is generally no use made of the function of the device nor is there aware- ness of device inputs or outputs that exist physically but which have not been listed in the image. The only check that is usually carried out on an image—apart from syntax checks—is a test to discover any listed device output that does not subsequently become either a primary output or an input to another device. Such outputs are called "floating" outputs and are listed by the image compiler.

To help reduce unnecessary probing, a number of strategies are used, some of which are illustrated in Figures 7.2 and 7.3.

The device in Figure 7.2, the SN 7404 hex inverter pack, should be imaged as six different devices, each at the same location, to remove unnecessary probing. If the probe is tracing back through the circuit and reaches one of the outputs, pin 10 say, then the only input it should probe is pin 11. If the chip is imaged as a single device with six inputs and six outputs, then the guided-probe algorithm would direct the operator to

Imaged as :

Device Name	Input Pins	Output Pins
U6	1	2
U6	3	4
U6	5	6
U6	9	8
U6	11	10
U6	13	12

and not :

U6	1,3,5,9 11,13	2,4,6,8, 10,12

FIGURE 7.2 Image for SN7404 Hex Inverter

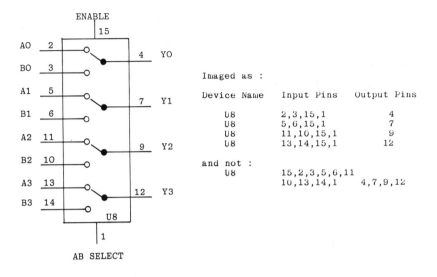

FIGURE 7.3 Image for SN74157 Multiplexer

probe all the input pins listed in the image up to pin 11. That is, the operator would be directed to probe pin 1, pin 3, pin 5, pin 9 and, eventually, pin 11. The first four probes in this sequence are unnecessary.

The second multi-output device, shown in Figure 7.3, is the SN 74157 multiplexer. For this device, each output is a function of two data inputs and the two common control lines—the ABSELECT and CHIP ENABLE controls. The device is imaged therefore as four different devices but the two common control lines are included in each of the four devices. The guided-probe algorithm will now instruct the operator to probe not only the two data inputs that directly affect a particular output but, if the two data inputs are found to be fault-free, then the operator will be instructed to probe either one or both control lines as well, depending on which line is faulty or is transmitting a fault-effect.

These two examples show how input-pin probing can be reduced around specific types of multi-output devices. Another aspect of this problem will now be considered.

The probing sequence on the input pins of a device whose output has been found to be faulty usually follows the order in which the pins have been listed. This ordering can therefore be instrumental in helping to reduce the length of a probing sequence. If some of the inputs on a particular device are part of a long and tortuous feedback/feedforward path and other inputs are not, then the feedback/feedforward inputs should be imaged after the other inputs.

A case in point is illustrated by the circuit used earlier in Chapter 3. (See Figure 3.14, page 61)

In the center of this diagram is a 2-input NAND gate, device U11, with inputs on pins 8 and 9 and output on pin 10. Initially, at least, it does not seem to matter whether pin 9 is imaged before pin 8 or vice versa. A closer inspection of the circuit, however, reveals that pin 8 goes almost directly back to two primary inputs, A14 and A18, whereas pin 9 is part of a feedback loop that starts at the gate in question, goes forward through the U6 J-K flip-flop and thence to the U2 decoder and back around to the gate again. Consider now the probing sequence for a fault on, say, one of the two primary inputs A14 and A18. Assume that the fault affects U11 pin 8 and therefore the devices in the loop, that is, U6, U2, and eventually, U11 pin 9. When the probing sequence returns to the U11 NAND gate, it will direct the operator to probe the first input in the image. If this node is U11 pin 8, then the sequence will continue back towards the primary inputs and the fault will be diagnosed fairly rapidly and with no unnecessary probing. If the first imaged node is U11 pin 9, however, then the probing algorithm may direct the operator to probe right around the loop; there is now considerable risk of an incorrect diagnosis.

These examples were chosen to illustrate how unnecessary probing can arise and possibly be removed. The other requirement for the image is that it should assist the process of accurate diagnosis. An example of how inaccurate diagnosis can arise has been presented in the previous chapter. (See Fig. 6.17.) The problem here was caused by the unbuffered nature of the outputs of the device. Other diagnostic problems also exist, particularly with regard to discrete components such as pull-up resistors. There are techniques to overcome these problems but they tend to be specific to a particular tester and are not described here.

Writing the image is not a trivial exercise and can have a profound effect on the final ability of the test program and the tester to locate accurately the source of a fault.

7.5 Stage 5 Board Set-Up and Initialization

Board set-up is the process of writing the test-program code to:

  (i) identify the program,
 (ii) print out instructions to the operator on how to set the board on the tester,
(iii) establish the EQUATE table,
 (iv) specify the power supply requirement,
  (v) specify the driver-sensor reference values,
 (vi) specify the STROBE value,
(vii) specify any other specific mode of the tester, e.g., SKEW, BROADSIDE, etc.

These requirements have already been discussed elsewhere (Chapter 1, Section 1.9).

Initialization has also been mentioned earlier in this chapter (section 7.2). The only additional comment to make at this point is that the process is often complicated unnecessarily by certain design practices. This aspect will be pursued more thoroughly in Chapter 9.

7.6 Stage 6 Write and Debug Test Program

Chapters 2 and 3 covered the main concepts and strategies for writing tests. The following general observations should also be made.

First, it is neither necessary nor desirable to develop the program line by line while sitting at the tester or at a terminal with a logic simulator. Thinking time with ''off-line'' preparation of the code is as important as ''on-line'' development.

Second, the general principles of writing computer software are just as applicable to a test program as to any other form of application software. That is, software should be well structured, easy to read, employ comment statements, and make use of left margin indentation as far as possible. Using these techniques will simplify the process of modifying the program sometime, should this become necessary.

Third, the test programmer should beware of the ''perfection at

infinity'' syndrome. Progress made in writing the test program is natur-
ally accompanied by a better understanding of how the circuit works and
how, perhaps, it could have been tested better. This produces the desire
to go back and modify the earlier code. The decision as to whether this is
worthwhile is generally left to the programmer and, in the end, is a
balance between the desire to produce an elegant program and the
practical economics of completing the program. The latter usually wins.

With these comments in mind, the test programmer now embarks
on the detailed construction of the program. To do this requires an
intimate knowledge of the test-programming language statements and of
how these statements are interpreted by the tester. In broad terms, the
language will contain the following types of statement.

(i) Identification of primary inputs and primary outputs

These statements will specify certain symbolic names as driven
inputs and sensed outputs. The facility may also exist to specify nominal
pull-up or pull-down currents on certain outputs. These currents are
supplied by the driver/sensor pin electronics and are useful for unterminated TTL open-collector or ECL open-emitter devices. The pull-up,
pull-down facility can also be used to test the tristate status of a
tristatable device.

(ii) Driving and sensing statements

These are the basic statements for specifying primary inputs to be
driven high or low, and for stating the expected high/low response of
primary outputs. Also in this context, it is often required to monitor
some outputs but to neglect others. The default option is usually to
monitor all pins declared to be primary outputs and then to neglect
specific outputs as and when the need arises. The monitor statement is
used to reinstate neglected outputs.

(iii) Generation of data sequences

Often there is need for regular data sequences on single inputs or
groups of inputs. Examples are clock sequences of a specified length,
binary up-count sequences (on the address lines of a RAM, for exam-

ple), Gray-code up-count to restrict data changes to single-bit changes, etc.

Specific statements usually exist to generate sequences such as these. In addition, general-purpose DO-loop or subroutine facilities exist to allow the programmer to create specialized sequences.

(iv) Conditional pin-state testing

Test-and-branch instructions have some value both for pin-state testing and for initialization or synchronization purposes. (A statement of this form could be used to recognize the all-0s state of the self-starting counter discussed in Chapter 3). Often both branch-if-fail and branch-if-pass statements are provided.

(v) Subroutine facilities

As with other programming languages, a subroutine facility is useful. Of particular value, for example, is a general initialization routine that can be called on at the beginning of each major segment of the test program.

(vi) Documentation and user-interaction

Useful facilities here include print, comment, and display statements for documenting the code and for displaying information to the user.

(vii) Debugging aids

Debugging a test program can be a time-consuming and frustrating exercise. Principally, the programmer must discover why either the board or the test program did not behave as expected. The reasons for discrepancies can be many and varied and facilities such as "stop-and-hold," "single-step," or program loop (use of GO TO statement) become invaluable for finding the cause of the problem. Debugging a program in this way usually involves the use of external aids such as logic probes (single point and multi-chip), digital multi-meters, and oscilloscopes. Figure 7.4 illustrates the use of a logic probe.

FIGURE 7.4 Using a Logic Probe (Courtesy Hewlett-Packard)

7.7 Stage 7 Fault-Cover Evaluation

There are two methods for evaluating the fault cover of the test patterns. The first is to use a logic simulator, the second to insert real faults on the board. Both these methods have been described in some detail in Chapter 4.

The fault list is usually derived from the image listing but, normally, faults are inserted only on the outputs of imaged logic devices and not on other imaged components such as pull-up resistors.

Running the program repeatedly can be achieved by use of a GO TO statement just before the end of the program. Alternatively, most testers will have command facilities to ''restart-on-end'' or ''stop-on-fault.'' (These facilities are also useful for debugging test programs.)

Fault insertion onto a board containing TTL devices can be carried out by means of a single wire, earthed at one end and clipped onto the device output pin at the other. Stuck-at-1 faults can also be inserted onto CMOS and ECL boards via a lead and current-limiting resistor connected to the positive or negative power supply lines. Faults that are not detected are marked for further attention and extra tests generated by appropriate means.

7.8 Stage 8 Learning and Verifying Signatures

If the diagnosis is to be based on the use of a guided probe, then the fault-free reference signatures (CRC or Transition Count) must be determined either from the board itself or from the simulated node value results.

Learning signatures from the board requires the availability of a known-good-board. This is one good reason for checking each step of the program during the main development stage (Stage 6). The tester works from the image database, instructing the programmer to place the probe on an internal node. When positioned correctly, the board is stimulated by the test patterns and the nodal response collected and fed into either the CRC register or the transition counter. The computed nodal signature is then added to the image and the next probe point indicated.

One problem with this procedure is that the programmer may inadvertently place the probe on the wrong node. Consequently, the signature that is learned and stored will almost certainly be incorrect. Fortunately, this occurrence is usually trapped and corrected by the next phase—signature verification.

Normally, signatures are learnt only on imaged outputs. This means that a verification stage is possible by observing the signature on an imaged input connected to this output and comparing the two signatures. The connectivity between the output of one device and the input to another is contained in the image; it is a simple matter for the tester to carry out this comparison using the image as a database and instructing the programmer to position the probe accordingly. Verifying signatures, therefore, is a process similar to the learning process, and inconsistencies can be flagged for attention.

The two most common causes of an inconsistency are:

(i) the programmer has misprobed either during the learning phase or during the verify phase, or

(ii) there is an error in the image—that is, the connectivity defined in the image does not agree with what physically exists on the board or what is defined by the circuit schematic.

Whatever the cause, the problem must be traced and rectified. If it is a misprobe during the learning phase resulting in an incorrect signature being learned, then the programmer will need to return to this stage and relearn the signature for this node. If the misprobe occurs during the verify stage then no further correction need take place other than to probe the right node and try again.

If the problem is more devious and is eventually traced back to a problem in the image, then the programmer will need to correct both the source code and object code versions. It is not usually necessary to learn all the signatures again—only those on the outputs that have been changed.

7.9 Stage 9 Checking Fault Location

Checking the fault-location properties of the program is achieved by inserting a fault and running the tests. The board should fail and enter the diagnostic procedure, e.g., fault dictionary with or without guided probe, guided probe alone, dual guided probe (for comparative testers), or programmed diagnosis (pin-state testing). Whatever the procedure, the following points are relevant.

First, the question of where to put the faults. The programmer is looking for faults that present a real challenge to the program from the location point of view. These are faults that lie within complex feedback paths or that are near the primary inputs of the board and that propagate along many sensitive paths. In general, therefore, the selection criterion is based on the anticipated difficulty of locating a particular fault.

The second comment is that, occasionally, the diagnosis is either ambiguous or clearly incorrect. The most common reason for this is inadequate imaging and this is why there was so much emphasis on the imaging stage earlier in the chapter. The cause may also be traced back to the test program.

Incorrect diagnoses should be corrected. This may mean amendments either to the image or to the program, or to both. Modifications to the program will almost certainly require a re-run through the signature-learning and signature-verifying stages, either partially or completely depending on the severity of the modifications. Any modifications to the outputs in the image will also require a return to the signature learn-and-verify stages.

In general, therefore, checking the fault-location properties may turn out to be a major exercise, just when the end appeared to be in sight.

7.10 Stage 10 Documentation

The program is now complete. The only job left is to ensure that all documentation, either of an archival nature or necessary to support the use of the program in the field, has been completed. Most of the documentation will be archival but it is important that the documents be filed in case modifications are necessary sometime in the future. Figure 7.5 lists the support documentation requirements.

7.11 Using the Program

Finally in this chapter, we take a brief look at some of the problems of using a test program as seen from the end-user's point of view. In

1. Test program notes, test plan and imaging notes.

2. wiring specification for board-tester interface.

3. Test performance summary (fault-cover with detected and undetected faults, faults selected to demonstrate location properties, difficulties and unresolved problems).

4. Listings and printouts: test program, image source, signatures learnt, signatures verified, marked-up fault-list, guided probe diagnoses and other printouts produced by a simulator.

5. Reconstructed circuit diagram plus comment on any differences between board and schematic.

6. magnetic tape or floppy disc copies of test program and image.

7. Test set-up and run instructions (tester type and operating system, tape/floppy number and file locations, any physical changes to the board, any special test clips, any ancilliary equipment required, e.g. oscilloscope, run instructions).

FIGURE 7.5 Test Program Support Documentation

general, problems of use can be categorized as: those due to the operator; those due to the test program; and those due to the tester.

7.11.1 Problems Caused by the Operator

Most operator problems are caused by inattention to detail when setting up the tester and loading the test program, image, or board. The following checklist indicates the sort of things that can go wrong.

    (i) Is the tester switched on?

   (ii) Is the correct version of the test program (or image) loaded? (Engineering changes usually require different versions of programs.)

  (iii) Is the correct version of the tester operating system loaded?

  (iv) Is the board-under-test the correct board?

   (v) Is the board justified, aligned, and oriented correctly?

  (vi) Some testers have zero-insertion-force connectors. Are such connectors closed properly?

 (vii) Is the interface connector the correct one and is it inserted correctly?

(viii) Is the tester mode correct and are any special switches or other facilities set correctly?

  (ix) Is the guided probe in the correct socket (and not in the +5v supply socket for a logic probe, say)?

Most of these problems are eliminated by visual inspection.

7.11.2 Problems Caused by the Test Program or Image

The three main problems here are:

  (i) the board passes sometimes, but fails occasionally;

 (ii) fault diagnosis is either ambiguous or even incorrect;

(iii) the board passes consistently on the tester but fails in the system.

Intermittent failures can be due either to genuine fault conditions that cause intermittent behavior (e.g., crosstalk, component degradation) or to program faults (e.g., marginal STROBE time, imprecise initialization, etc.). In either event, tracing the cause of the problem can be time-consuming.

Ambiguous or incorrect diagnosis can also stem from many causes, some of which have already been discussed, such as unbuffered outputs, presence of feedback loops, and discrete components omitted from the image. Bus-structured designs can also create problems; this aspect is discussed further in the following chapter.

Finally, the no-fault-found phenomena have been mentioned in the discussion on the effect of inadequate fault cover. Sometimes the board is genuinely fault-free but fails in the system because of a mating connector problem. Whatever the reason, solving the no-fault-found problem can prove to be very exasperating.

7.11.3 Problems Caused by the Tester

The tester is just as prone to fault conditions as any other digital system and these are usually detected by running a special self-test program. The most vulnerable part of the tester is at the interface to the board-under-test, i.e., at the operational amplifiers at the driver/sensor pins. Other vulnerable areas include the programmable power supplies and the mechanical interfaces for tape-cassette and floppy-disc systems, in short, any part of the system that can be subjected to mis-use.

Chapter 8
Testing Boards Containing LSI/VLSI Devices

The advent of LSI and VLSI devices has thrown an additional strain on the processes of testing digital boards. Chapter 8 discusses the new problems and shows how some, at least, of them can be overcome. As a starting point, we make the following assumptions about the board and our approach to testing it.

(i) The general strategy for generating tests is as described in Chapter 3, section 3.6, but with more emphasis on checking each device against some form of a functional checkout sequence.

(ii) It is still thought desirable to use a fault simulator to assist in test-pattern generation and possibly to evaluate fault cover.

(iii) LSI/VLSI devices can be classified into the following generic types:
Fixed memory (ROM, EAROM, PROM)
Read-write memory (RAM)
Microprocessor/Microcomputer (MP/MC)
Receiver/Transmitter (UART, USRT)
Controllers (DMA, Interrupt, Disc, I/O, etc.)
Buffers, line drivers, and multiplexers

(iv) Information transfer (address, data, and control) is largely by way of a bus system.

Figure 8.1 shows the general form of a PCB containing LSI/VLSI devices. The special case of memory boards (mostly containing RAM) will be deferred until later in this chapter.

We will now reconsider the three major activities of testing and indicate how the presence of LSI/VLSI devices introduces new problems.

8.1 Test-Pattern Generation Problems

8.1.1 Device Complexity

Understanding how devices react to particular stimuli is fundamental to test-pattern generation and, at this level, can become a major problem, particularly for microprocessor/microcomputer devices and also for certain controller devices. The problem is compounded also by:

    (i) data sheets that are either ambiguous, incomplete, or even incorrect;

   (ii) differences between prime-source devices and those supplied by second-source manufacturers;

  (iii) the variety and number of new devices that are becoming available.

An example of an incomplete data sheet occurs with the Intel 8253 Programmable Interval Timer. This chip contains three independent

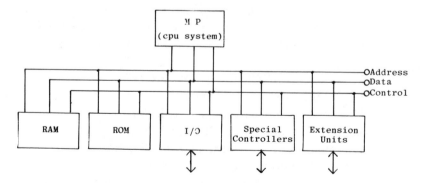

FIGURE 8.1 General Structure of a Bus-Structured Microcomputer Board

16-bit counters whose operating mode (binary count, BCD count, square-wave generator, etc.) is determined by a control word held in an internal register. Each counter has an external CLOCK input that can be inhibited by a separate external GATE control input. The counter will only count if GATE is high. If GATE is low, the counter output is held high.

In mode 2 (divide by n mode, where n is variable and preloaded) the GATE should go high $100nS$ before the positive edge of the CLOCK (minimum set-up time). If this time is violated, then what we would hope is that the output remains high until the next positive edge on CLOCK, after which the divide-by-n count can commence. What actually happens is that the output goes low and stays low until the end of the n count cycles. (See Figure 8.2.)

An example of prime-source, second-source differences occurs with the Texas Instrument 6011 UART device, second-sourced by General Instrument Microelectronics as the AY-5-1013 UART. Parallel input data is loaded into the Transmitter Buffer Register (TBR) when the TBRLoad signal is low. Transfer from the TBR into the main transmitter register (TR) occurs as TBRL goes high. During the load and transfer period, TBREmpty flag is set low to indicate that the TBR is not able to accept new data. For the TI device, TBRE goes low at the start of the load cycle (TBRL going low) whereas for the GI version, TBRE goes low at the end of the load cycle (TBRL going high.) (See Figure 8.3.)

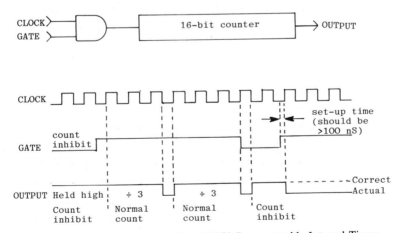

FIGURE 8.2 Mode 2 Operation of Intel 8253 Programmable Interval Timer

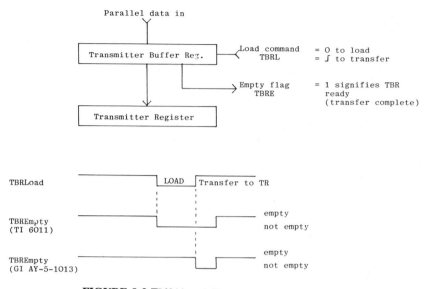

FIGURE 8.3 TI6011 and GI AY-5-1013 Differences

8.1.2 Device Accessibility

The fact that information flow—addresses, control, and data—is largely organized via a bus structure is both an advantage and a disadvantage. The advantage is that it is relatively easy to determine how to control and observe a device; the disadvantage is that either controllability or observability may be through other complex devices. It is necessary not only to understand how individual devices work but also to understand how devices of different type interact.

8.1.3 Volume of Test-Pattern Data

As we shall see, the amount of test-pattern data required to test a PCB as complex as a single-board computer or a RAM store board can become quite large. It is important, therefore, that the generation of this data be carried out in a structured manner and that, if necessary, particular sections of the code can be applied in isolation from other sections. This approach is important for program development and debugging purposes.

8.1.4 Failure Mechanisms

A major assumption in testing boards containing SSI/MSI devices is that any failure mechanism internal to a device will manifest itself as a solid stuck-at fault at either the device inputs or outputs. It was sufficient therefore to provide test patterns against a nodal stuck-at fault list. In practice, however, we recognize that certain failure mechanisms are not necessarily modelled by the single nodal stuck-at model; practical test programs contain extra tests to detect bridging faults (i.e., unidirectional multiple faults) between specific circuit nodes.

For LSI and VLSI devices, this assumption is no longer valid. There are internal failures which do not necessarily manifest themselves as solid stuck-at faults on the device input or output pins. Specific strategies have been developed for these failures. These strategies will be discussed more fully when we consider how to test specific types of device, but the sort of failures that occur are bit drop-out in MOS RAMS caused by charge leakage or, in more recent devices, alpha-particle radiation. The charge-leakage problem occurs if a particular pattern of 0's and 1's is present in the RAM and is known as the "pattern sensitivity" problem. (See Figure 8.4.)

Consider the situation as shown, i.e., cell i storing a logic 1

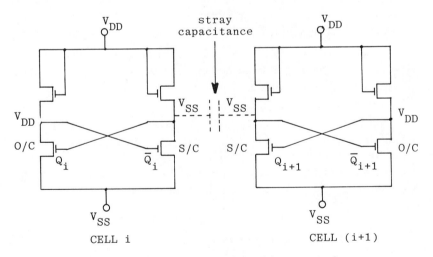

Row and column select lines and bit lines not drawn.

FIGURE 8.4 Interaction Between MOS RAM Cells

(right-most transistor $\overline{Q}_i$ acting as a short-circuit) and cell $(i + 1)$ storing a logic 0 (right-most transistor $\overline{Q}_{i + 1}$ acting as an open circuit). If the state of cell i is changed, i.e., if $\overline{Q}_i$ drain goes to V_{DD}, then it may be possible for the change of $\overline{Q}_i$ drain to couple through to $Q_{i + 1}$ via stray capacitance and hence cause cell $(i + 1)$ to change state erroneously. This is an example of pattern sensitivity caused by adjacent-cell disturbance.

Alpha-particle radiation also occurs in 64K MOS RAMs and originates from trace elements of thorium and uranium in the ceramic packaging material. The energy level can be up to almost 9 million electron-volts (ev); an alpha-particle of 5 million ev can penetrate silicon to a depth of 25 μm, producing up to 1.4 million electron-hole pairs. Depending on where the penetration occurs, this increase in charge can cause erroneous single-bit changes of either 0 to 1 or 1 to 0. If penetration occurs on the refresh reference cell during a refresh cycle, multiple errors can occur as a result of a change in the value in the reference cell.

Failures caused by pattern sensitivity and alpha-particle radiation are known generically as "soft" errors, that is, they are not necessarily repeatable and their occurrence is generally random.

Another source of error arises from internal failures that affect propagation parameters and give rise to timing interaction faults. An example of this is a missing or defective Schottky diode in a Schottky TTL gate. The gate will still function correctly but its propagation delay will be increased. The only way to detect the presence of faults of this type is to test both individual devices and device-device interaction (timing responses or cumulative propagation delays) at the speed at which they are intended to operate.

In general, therefore, it is not sufficient to test the board against a nodal stuck-at standard. The test programmer must provide tests for other forms of failure if they are known or suspected. In this respect, therefore, the evaluation of the final set of test patterns is no longer a straightforward task. This is the subject of the next section.

8.2 Test-Pattern Evaluation Problems

Traditionally, as we have seen, logic simulators have been used to measure the effectiveness of a set of input stimuli by determining the fault-cover in terms of nodal stuck-at faults or other well-defined logical fault effects. The extension of this use into boards containing LSI/VLSI devices becomes complicated for the following reasons.

(i) Modelling Level and Faulty Response

We have already remarked on the inadequacies of data sheets for certain devices, particularly recent devices. Even if the operation of the device is well understood, there is still the question of modelling level. Should the device be modelled in terms of its equivalent logical circuit (gates, shift-registers, random-access memory, internal busses, etc.) or should it be modelled at a higher functional level? The equivalent logic circuit drawn in the specification sheet for the device may be useful to the application programmer but incomplete for the test programmer.

Even supposing we are able to develop a suitable model to represent the fault-free behavior of the device, we still need to incorporate information as to how the device will react to faulty input stimuli (if the device is acting as a fault transmitter) or to a particular internal failure mechanism (if the device is acting as a fault generator). Both these requirements can create further problems. If the input stimuli are such as to cause the device to go into an "illegal" state* then the model may not produce the correct output response. If the failure is internal, the precise response of the device to any particular sequence of input stimuli or internal instructions may be unknown.

(ii) Other Modelling Problems

Other features that are common to bus-structured PCBs are open-collector or tristate devices together with bidirectional busses. Logic simulators have to be capable of modelling the electrical status of these devices—logic 1, logic 0, or high impedance (open-circuit)—and also whether information is passing from device A to device B or vice versa. As before, the simulator should also be able to handle the outcome of, say, two devices that access the bus simultaneously—one correctly and

---

*An example of an illegal state occurs with the Intel 8253 Programmable Internal Timer. The control word for this device contains a 2-bit field (SC1, SC0) to identify which of the three counters is to be selected. SC1 = SC0 = 0 selects counter 0; SC1 = 0, SC0 = 1 selects counter 1; and SC1 = 1, SC0 = 0 selects counter 2. The other possibility of SC1 = SC0 = 1 is illegal and the outcome of this assignment is unspecified.

the other because of a fault. Some simulators are not equipped to model tristate values and bidirectional busses directly but do allow "work-around" solutions. An example of this is shown in Figures 8.5 and 8.6.

Figure 8.5 shows two devices with tristate outputs connected to a bidirectional bus. Figure 8.6 shows one solution to how the behavior can be modelled. The objection to this solution is as follows:

If $CS(A) = 0$ (fault-free) and $CS(B) = 0$ (fault generator or fault transmitter), what is the outcome if $DO(A) = 0$ and $DO(B) = 1$? The simulator will say BUS $= 1$, but in practice the result is indeterminate.

8.3 Test-Application and Fault-Finding Problems

8.3.1 Tester Considerations

The main considerations with regard to choice of tester are test-application rates and driver-sensor turnaround. Test-application rates could not be of the order of 1 million patterns per second for real-time tests, and the use of bidirectional busses will require driver/sensor pins to be switched from driver mode to sensor mode, and back, at real-time

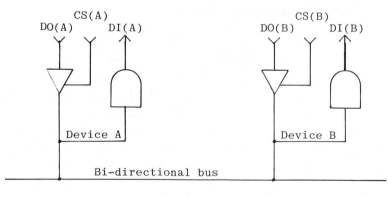

```
DO(A), DO(B) are data outputs (onto the bus)
CS(A), CS(B) are chip-select (tristate) control lines
DI(A), DI(B) are data inputs (from the bus)
```

FIGURE 8.5 Tristate Devices on a Bidirectional Bus

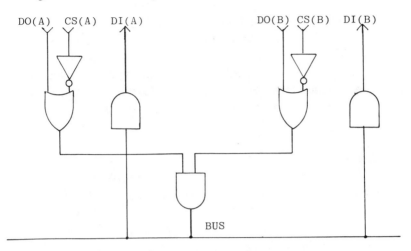

Workaround solution.   Chip select (CS) lines take binary
values (0 to model high impedance state, 1 to output data
onto bus).

FIGURE 8.6 Simulator Model

speeds for some of the time at least. If it is not possible to satisfy either of
these requirements, then the real-time tests will have to be omitted with
accompanying risk of failure to detect certain error conditions.

For certain devices, there may also be a minimum speed of operation. The Intel 8080A microprocessor, with its minimum operating
speed of 500 kHz (2 μs clock), is a case in point.

Another consideration is whether it is necessary to synchronize the
tester to the circuit (rather than the more usual situation of synchronizing
the circuit to the tester). For example, the Intel 8224 Clock Generator
and Driver contains an internal reference oscillator whose frequency is
determined by an external crystal (8801 device). It is not possible for the
tester to take the place of the internal oscillator. This may mean that, for
certain real-speed tests, the tester will have to synchronize to the
oscillator output available from the 8224 chip. An enlightened logic
designer will make this output available at the board edge connector.

Finally, there should be some mechanism for masking, for signature purposes, the collection of binary data from a node that may assume
a high-impedance tristate condition. Unless special features are built
into the probe (as in the probe for the HP 5004A Signature Analyser

discussed in Chapter 9), the logic value of a node when it is in its high impedance state is generally indeterminate. CRC signatures, therefore, can easily become corrupted or be inconsistent if there is no possibility of preventing data collection during the high-impedance phases of a node. One solution to the problem is to make use of the monitor/neglect features of the test-programming language (as discussed in Chapter 7).

8.3.2 Fault Location

A problem peculiar to bus-structured systems is the problem of precise identification of a faulty device whose only visibility to the tester is via the bus, e.g., ROM, RAM, MP, I/O devices, etc. The problem is to locate the fault "beyond the node."

Essentially, the problem is that if the outputs of all devices that have access to a bus are supposedly at tristate status and yet the bus cannot be pulled high (or low), then this suggests that one of the devices at least is faulty in some way. The problem is—which device?

To illustrate this, consider the situation shown in Figure 8.7, in which one of the RAM chip-select lines contains a fault, causing data to be loaded onto the bidirectional data bus. The tester detects erroneous data on the bus and, by virtue of the way the test program has been written, directs the user to probe some other device first—say the serial I/O device. A standard voltage-sensing probe will identify the bus-side

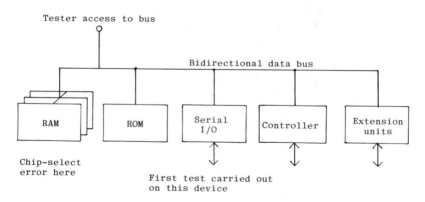

FIGURE 8.7 Fault Location "Beyond the Node"

pins of this device as faulty but the serial-data pins as fault-free. Diagnosis therefore is to the I/O device since it satisfies the criterion of the locating algorithm to ''find the device whose inputs are fault-free but whose outputs are faulty.''

It is simple enough to overcome this problem—include data transfer both into and out of the I/O device—but doing so creates another problem: the diagnosis will now become ambiguous and, in the limit, could be ambiguous to all devices on the bus. An associated problem is that for some devices, certain inputs are physically the same as the outputs. An example is the Motorola 6810 RAM device. The eight data input pins are physically the same as the eight data output pins, i.e., the pins are bidirectional. Effectively therefore, a feedback loop exists. As remarked in an earlier chapter, breaking fault-effect propagation paths around a feedback loop is difficult enough without the added complication of physically identical inputs and outputs.

One solution to the problem is to use another type of guided probe, namely, one that identifies the precise location of the source, or sink, of the current on the bus. Current-tracing probes are now available. In the hands of a skilled operator, these probes can be used to follow the actual printed-circuit track until the source of the problem is determined.

One particular form of current-sensing probe works on the principle of induced voltage from an alternating current source supplied either by the tester providing pulse stimuli to the board or by another special-purpose logic pulser probe. Figures 8.8. to 8.10 show how a current-tracing logic probe can be used to assist fault location for various types of fault, and Figure 8.11 shows an actual probe in use.

Another form of current-sensing probe, known as the ''electronic

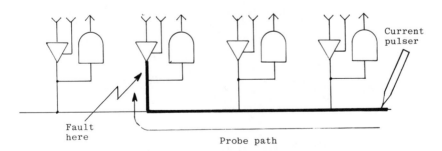

FIGURE 8.8 Locating A Fault on A Bus Line

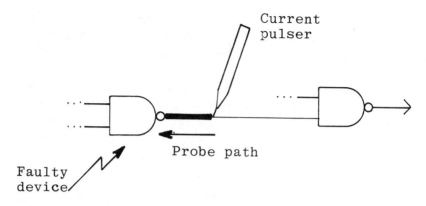

FIGURE 8.9 Isolating the Cause of a Nodal Fault

knife,'' also exists. Unlike the previous probe, the electronic knife is a three-point contact probe designed for use on the pins of an integrated-circuit chip. The mode of use is shown in Figure 8.12.

8.4 Testing Specific LSI/VLSI Devices

The following sections comment on particular strategies for testing specific LSI/VLSI devices. The strategies to be described are not neces-

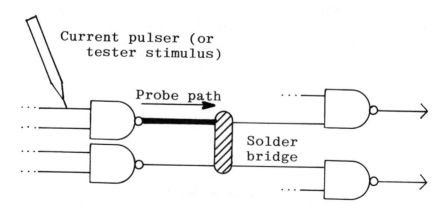

FIGURE 8.10 Locating a Solder Bridge

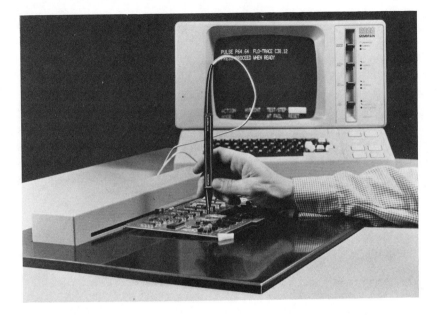

FIGURE 8.11 "Flo-Tracer" Current Sensing Probe (Courtesy Membrain)

sarily the only strategies for devices of this type; neither is it claimed that they are the best. The art of testing devices such as these, and boards that contain these devices, is continually changing. All we can do here is to indicate current trends.

In general, the aim is to provide a functional checkout sequence for each device on the board. This sequence is based on:

(i) an understanding of how the device works;
(ii) an understanding of how it might fail.

The format of the following sections is to discuss device testing strategies in general terms and, in the diagrams, to summarize their salient architectural features.

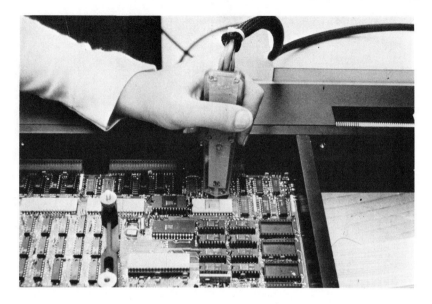

FIGURE 8.12 "Electronic Knife" Current-Sensing Probe (Courtesy Teradyne)

8.5 Fixed Memory (ROM, PROM, etc.)

Read-only memory devices are essentially look-up tables. (See Figure 8.13.)

Possible failure modes for this device are:

(i) incorrect addressing (decoder malfunction)
(ii) incorrect chip select
(iii) incorrect contents in the cell matrix

The simplest form of test is to carry out a full truth-table checkout, i.e., to sweep through all addresses and check the outputs against known data. In its simplest form therefore, the test program contains the input address together with the expected output response. Obviously, this can become a tedious exercise. A slight variation is to specify just the input

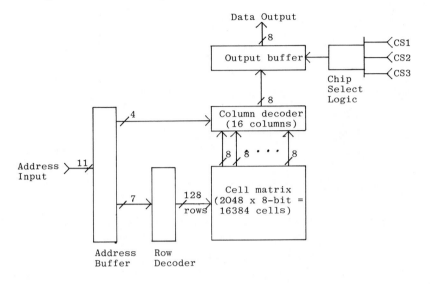

FIGURE 8.13 ROM Architecture

address and allow the tester to learn automatically the output response—either in pure binary form or in some compressed form such as a CRC signature. This is only feasible if the ROM used for learning is a known-good-ROM.

Alternatively, checksums can be precalculated and stored in the test program. A disadvantage with the simple form of checksum, which is really no more than an even-1 column parity test, is that it may miss faults that affect data in columns rather than rows. Such faults can either be caused by a real failure mechanism or by transcription errors. The following procedure, based on calculating a skew checksum, overcomes this.

Step 1. Set CSUM = 0

Step 2. Set Address A = all-0 address

Step 3. Rotate CSUM one position left

Step 4. CSUM = CSUM + contents (A)

Step 5. A = A + 1

Step 6. If A has exceeded last address, go to Step 7. Otherwise go to Step 3.

Step 7. Stop. CSUM contains the final checksum.

8.6 Random-Access Memory (RAM)

In principle, the RAM is a simple device, as shown in Figure 8.14.

Unfornately, both bipolar and MOS RAMs are susceptible to many different forms of failure and this has given rise to literally dozens of test strategies. The main problem has been to provide a set of tests that is not excessively long and yet covers the possible failures, particularly those that can be attributed to pattern sensitivity.

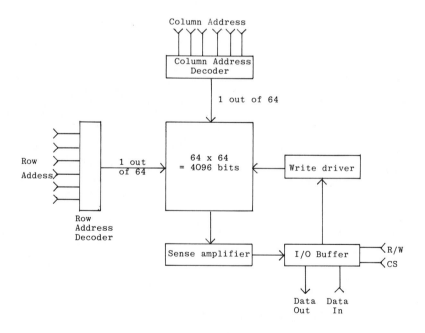

FIGURE 8.14 RAM Architecture

For MOS RAMs a list of possible failure modes will include:

(i) incorrect addressing (decoder malfunction)
(ii) multiple writing (usually caused by capacitative coupling)
(iii) pattern sensitivity (during either read or write operations)
(iv) refresh sensitivity in dynamic RAMs (failure caused by excessive charge leakage to retain data during the minimum refresh cycle time)
(v) slow access times (too much capacitative charge on output driver circuits)
(vi) sense amplifier recovery (sense amplifier saturates after recognizing a long sequence of 1s say, resulting in an excessive time to recognize a 0)
(vii) write recovery (slow to read after a write operation caused by saturated sense amplifier)

To illustrate something of the problem of testing RAMs, we will consider just five possible strategies—simple read/write; diagonal pattern and WALKPAT; MARCHPAT; GALPAT; and nearest-neighbor-disturb. For simplicity, we will assume a 4×4 16-cell square matrix where cells that are numerically adjacent are physically adjacent in rows except at the row ends. (See Figure 8.15.)

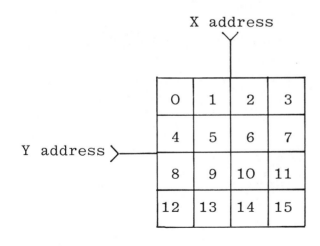

FIGURE 8.15 Layout of 16-Cell RAM

In practice, neither of these conditions are necessarily true. Cell matrices may be laid out on a rectangular basis; moreover, numerical adjacency does not guarantee physical adjacency.* To make matters worse, second-source suppliers of the same device type may make use of a completely different matrix geometry.

8.6.1 Simple Read/Write

The strategy here is as follows:

Step 1 Write a 1 to each location (cell) in the RAM

Step 2 Read each cell and check that it is 1

Step 3 Write a 0 to each cell

Step 4 Read each cell and check that it is 0

All this test proves is that at least one of the cells in the memory works. For example, if all the address lines were bridged to 0v (externally or internally) then the RAM would still pass the test.

8.6.2 Diagonal pattern and WALKPAT

The strategy for the diagonal pattern is shown in Figure 8.16.

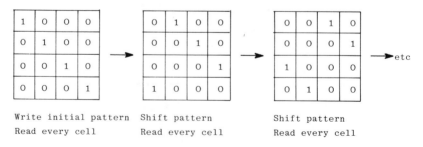

Write initial pattern   Shift pattern   Shift pattern
Read every cell   Read every cell   Read every cell

FIGURE 8.16 Diagonal Pattern

*Unravelling the exact relationship between cell addresses and their physical location is known by the glorious title of ''topological descrambling.''

The sequence will verify that the address decoders are functional. If multiple cells are selected (by virtue of capacitive coupling between cell address lines) then this will be detected as a 1 read back from a cell off the diagonal. Similarly, if a cell is totally inaccessible, i.e., is always wrongly addressed, then this will show up eventually either as a 0 in the diagonal of 1's or as a 1 in the background of 0.

A variation of the diagonal pattern is to progress a single 1 through the matrix rather than a diagonal of 1's, reading the full matrix after each progression. The variation is called the "walking ones" pattern or WALKPAT. (See Figure 8.17.) Obviously WALKPAT will take longer (why?)* but it is more effective against slow recovery of the sense amplifier.

8.6.3 Marching Patterns

Figure 8.18 illustrates the mechanism of the "marching-one" pattern.

The marching-one and its inverse, the marching-zero, are alternative patterns for detecting failures of the address decoders but, because of the limited readout, they may miss certain multiple addressing possibilities.

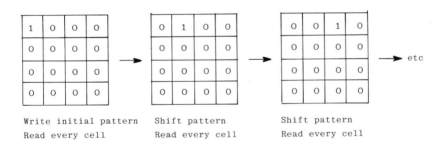

Write initial pattern   Shift pattern   Shift pattern
Read every cell   Read every cell   Read every cell

FIGURE 8.17 WALKPAT

*The number of WRITE operations remains the same but the number of READ operations increases. For the diagonal pattern, the total contents of the memory are changed 5 times producing, $5 \times 16 = 80$ READ operations, whereas for WALKPAT the total contents are changed 17 times producing $17 \times 16 = 272$ READ operations.

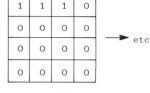

1	O	O	O
O	O	O	O
O	O	O	O
O	O	O	O

→

1	1	O	O
O	O	O	O
O	O	O	O
O	O	O	O

→

1	1	1	O
O	O	O	O
O	O	O	O
O	O	O	O

→ etc

Write initial pattern   Change pattern   Change pattern
Read back 1   Read all 1 cells   Read all 1 cells

FIGURE 8.18 Marching-One Pattern

8.6.4 Galloping Patterns

There are many variations of the galloping pattern techniques, known generically as GALPATs. Figure 8.19 illustrates one version of GAL-PAT.

    This figure shows a galloping-one sequence. The complement, galloping-zero, also exists and is usually applied after the galloping-one sequence. The prime objective of the GALPAT sequence is to identify read/write disturbance problems between a given cell and all other cells. Effectively therefore, the application of both forms of GALPAT (1s and 0s) should detect any pattern sensitivity problems.

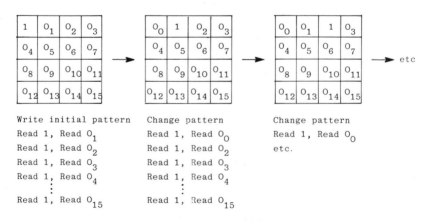

1	O_1	O_2	O_3
O_4	O_5	O_6	O_7
O_8	O_9	O_{10}	O_{11}
O_{12}	O_{13}	O_{14}	O_{15}

→

O_0	1	O_2	O_3
O_4	O_5	O_6	O_7
O_8	O_9	O_{10}	O_{11}
O_{12}	O_{13}	O_{14}	O_{15}

→

O_0	O_1	1	O_3
O_4	O_5	O_6	O_7
O_8	O_9	O_{10}	O_{11}
O_{12}	O_{13}	O_{14}	O_{15}

→ etc

Write initial pattern   Change pattern   Change pattern
Read 1, Read O_1 Read 1, Read O_0 Read 1, Read O_0
Read 1, Read O_2 Read 1, Read O_2 etc.
Read 1, Read O_3 Read 1, Read O_3
Read 1, Read O_4 Read 1, Read O_4
     ⋮   ⋮
Read 1, Read O_{15} Read 1, Read O_{15}

FIGURE 8.19 GALPAT (Galloping-One)

8.6.5 Nearest-Neighbor-Disturb Pattern

The nearest-neighbor-disturb (NND) pattern is a variation on GALPAT designed specifically for pattern sensitivity. The pattern is shown in Figure 8.20 for a galloping-one sequence.

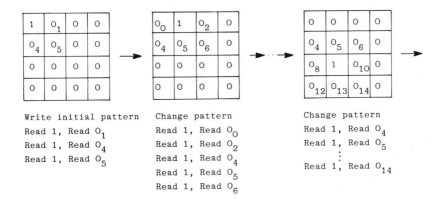

FIGURE 8.20 Nearest-Neighbour Disturb

Basically the strategy is to read only the cells that are physically adjacent to a given cell, rather than to read all cells. Variations of NND patterns include not only read/read but also read/write operations around the given cell in further efforts to detect pattern-sensitivity problems.

8.6.6 General Comments on RAM Testing

Not all RAMs exhibit all the failures listed above. This means that the test programmer must have a good understanding of the possible failures for the particular RAM with which he is faced. There is obviously no point in cycling through a complete GALPAT sequence (both 0s and 1s) if the device is not prone to pattern sensitivity. By the same token, there is every reason to apply a WALKPAT sequence if it is known that the device suffers from multiple addressing faults. Also, the amount of on-board RAM testing can be reduced quite significantly if it is known that each device has already been through a stringent goods-inward test or even through a separate characterization and burn-in test.* If this is so, then it is often sufficient to test the RAM to show that:

  (i) the data-in, data-out, and R/W lines are not s-a-1 or s-a-0;
 (ii) the chip enable is not s-a-1, s-a-0;
(iii) the address lines function correctly;
(iv) each memory cell can be set to 0 and 1.

For a PCB consisting mainly of RAM devices, i.e. a store board (See Figure 8.21), the following strategy has been found useful.

STEP 1 Initialize

Establish a background pattern of all-1s in every location of every RAM.

STEP 2 Data-out stuck-at test (bus test)

With no RAM selected, test data bus by pulling low and high. For open-collector, the rest state is high but it should be possible to pull the node low. For tristate, the rest state is indeterminate and can be pulled both high and low. Note that if the open-collector pull-up resistors are not actually present on the board, then they must be supplied either by the tester or by a special interface jig.

STEP 3 Nodal stuck-at and R/W stuck-at

Select RAM(1), CE(1) active, CE(2) − CE(8) inactive
Select a working address, A = 0000 say, with data in = 1111
R/W = Read, output should be 1111

---

*"Characterization" is the name given to the series of tests designed to test a device as fully as possible. These tests will include not only a functional checkout but also parametric tests, temperature-cycling tests, and voltage-supply tolerance tests. Implicit in the last two tests is an element of device burn-in. Devices that have been through a full characterization, therefore, are normally considered fault-free (although if the functional tests are of a limited nature, the fault-free property is not guaranteed.)

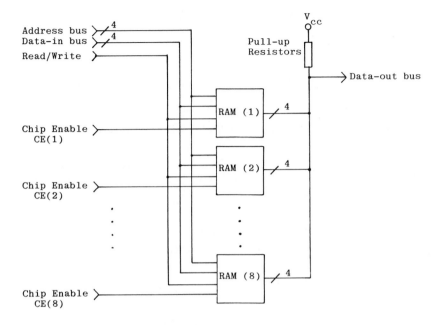

FIGURE 8.21 RAM Store-Board

If 1110, Data bit 0 is s-a-0
If 1101, Data bit 1 is s-a-0
If 1011, Data bit 2 is s-a-0
If 0111, Data bit 3 is s-a-0

Change data-in to 0000 but maintain R/W in Read mode
Output should still = 1111. If 0000, R/W is stuck-at-write
Change R/W => Write, Output should go to 0000

If 1111, R/W is stuck-at-read
If 0001, Data bit 0 is s-a-1, etc.

Restore to initialized condition
Repeat for other RAM chips

Note: This test assumes that the device is "transparent" when R/W = Write, i.e., data-in is transferred into the memory cells as R/W goes to the write level and is therefore available on the data-out lines for the duration of the write level. If this is not the case, then the procedure requires a slight modification.

STEP 4 Chip Enable Stuck-At Test

Initialize as before (Step 1)
Select a working address, A = 1111 say
Select RAM(1) and write 0000 into A
Read data-out. Should be 0000. If 1111, CE(1) is stuck-at-inhibit
Deselect RAM(1), select RAM(2)
Read data-out. Should be 1111. If 0000, CE(1) is stuck-at-enable
Select RAM(1) again, write 1111 back into A

Note that if both RAMs are selected, one correctly and the other incorrectly, it is assumed that the logic 0 on the data-out line will dominate the bus, i.e., the bus will go low.

Repeat test with RAM(2), testing against RAM(3)
                  RAM(3), testing against RAM(4), etc.

STEP 5 Address stuck-at faults

Starting from initialized values:
Select RAM(1), select working address A = 1111
Write 0000 into A
Read contents of addresses 1110, 1101, 1011, 0111
Output should be 1111
0000 from any of the four addresses indicates:

    (i) an address line s-a-1, i.e., data 0000 went to correct 1111
         address, or
    (ii) an address line s-a-0, i.e., data 0000 went to the wrong location
         to start with

Repeat for RAM(2)–RAM(8)

STEP 6 Address bridging faults

Assumption : wired-AND, i.e., 01 => 00, not 11
Initialize to all-1s
Select RAM(1)
Select working address A = 0000
Write 0000 to A
Read contents of 1000, 0100, 0010, 0001

Output should be 1111. A wired-AND bridging fault between two or more address-line inputs will pull address to 0000, resulting in 0000 output.

Before we leave memory testing, we will make one final comment. In practice, the difficulty with bus-structured boards that contain many memory devices is not how they should be tested, but how the diagnosis can be achieved without undue probing. Once the correct set of device tests has been constructed and programmed, the same tests can be repeated for all other devices of this type on the board. When one of the devices fails and corrupts the data on the bus line, the problem becomes one of locating the fault beyond the node. Solutions to this problem do exist, based on the use of programmed messages, but they tend to be tester-specific and therefore are not described in this book.

8.7 Microprocessor Devices

Microprocessor devices are currently the most complex devices to test. Figure 8.22 shows the salient architectural features of a microprocessor chip.

These features are explained briefly:

Program counter : contains address of next instruction
Flag register : individual internal or external conditions
Stack pointer : pointer to memory address used to store present status in the event of an interrupt or subroutine call
Index register : a register whose value is used to calculate operand addresses
Accumulator : immediate temporary storage area for operands
ALU : Arithmetic and Logic Unit
Register files : alternative temporary storage area (scratchpad)
Instruction Reg. : holds current instruction
Instruction Decoder : decodes current instruction and produces micro-instructions
Timing Unit : basic clock generator
Address Buffer : main memory address or peripheral address
Data Buffer : bidirectional data-in, data-out

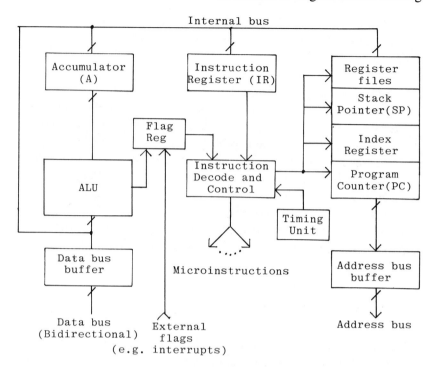

FIGURE 8.22 Microprocessor Architecture

Failure modes are numerous and, in some cases, unpredictable. Key failures modes, however, are:

(i) Internal single and multiple stuck-at faults
(ii) Timing faults due to degraded propagation delays, sensitivity to data load, sense amplifier recovery, other unknown effects. (Generically known as "instruction sensitive" faults.)

Testing strategies vary enormously according to who is doing the testing (manufacturer or user) and what the application is (low or high cost-of-error factor). Essentially, however, the only way to test the microprocessor is to perform the best possible series of tests on each of its sub-functions with the best possible awareness of how the device might fail. An overriding consideration, however, is the cost of generating and applying these tests, including the learning-curve period for a new device. In certain circumstances, it is cheaper to discover and debug the problems through field experience. In other circumstances, such as real-time process control, failure in the field can be very expensive or can have a catastrophic effect. The amount of goods inward and board checkout, therefore, can vary over a wide range. To illustrate how a functional checkout based on architectural considerations might proceed, the next section describes a possible plan for the Intel 8080 microprocessor.

8.7.1 Architectural Checkout Sequence

Figure 8.23 summarizes the architecture of the 8080 microprocessor; Figure 8.24 shows the instruction set; and Figure 8.25 the checkout sequence.

The following comments apply to the checkout sequence.

PROGRAM COUNTER TEST. A precursor to this test is to place the NOP code on the data bus and to initialize the PC to 0. For the 8080, the PC is 16-bits. A full sequence of pattern changes results in a test time of $2^{16} \times n$ clock cycles, where n is the execution time of the NOP instruction ($= 4$ cycles). Obviously a check of this type is no guarantee of correct PC operation for other instructions.

SCRATCHPAD MEMORY TEST. What is happening here is that each of the six 8-bit general-purpose registers (B,C,D,E,H,L) is being tested for all 256 patterns by loading the pattern into the H and L registers, via the address bus into the PC, and then transferring from H and L to D and E and then to B and C. The transfer is then reversed and finally brought back out to the address bus through the PC.

STACK POINTER TEST. Similar to the scratchpad memory test, i.e., increment or decrement the register through all its states (2^{16} in this case). Access and verify is again through the PC.

As an alternative to a full 2^{16} sequences, the stack pointer can be tested for stuck-at faults and adjacent bridging by loading all-0s, all-1s, chequerboard, and inverse chequerboard only.

ALU and ACCUMULATOR TEST. The problem here is sheer volume of tests. The ALU works with two 8-bit operands, one in the accumulator and the other from a variety of source or destination

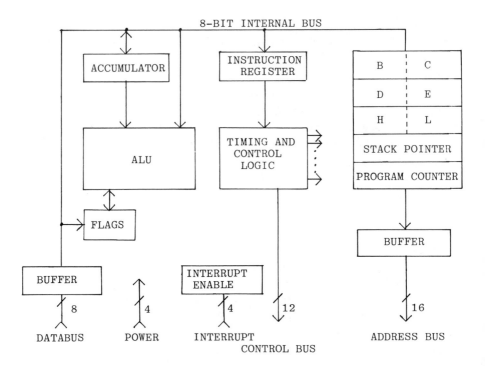

FLAGS : S Sign bit, set when msb = 1
        Z Zero bit, set when result of operation is zero
        P Set if parity is even
        C Carry bit
        AC Auxiliary carry (used with BCD operations)

INTERRUPT ENABLE (IE) set/cleared under EI and DI instructions.

H and L registers contain High and Low bytes of 16-bit pointer to memory.

Contents (H,L) is designated M.

PROCESSOR STATUS WORD (PSW) designates the ACCUMULATOR (A) + flag register.

Scratchpad cpu register-pairs are designated B (=B+C),D(=D+E),H(=H+L)

Instruction cycle times can vary according to whether cpu registers are selected (shorter time) or H is selected (longer time)

FIGURE 8.23 Architecture of the 8080 Microprocessor

192

8080 INSTRUCTION SET

Summary of Processor Instructions

Mnemonic	Description	D7	D6	D5	D4	D3	D2	D1	D0	Clock[2] Cycles
MOVE, LOAD, AND STORE										
MOVr1,r2	Move register to register	0	1	D	D	D	S	S	S	5
MOV M,r	Move register to memory	0	1	1	1	0	S	S	S	7
MOV r,M	Move memory to register	0	1	D	D	D	1	1	0	7
MVI r	Move immediate register	0	0	D	D	D	1	1	0	7
MVI M	Move immediate memory	0	0	1	1	0	1	1	0	10
LXI B	Load immediate register Pair B & C	0	0	0	0	0	0	0	1	10
LXI D	Load immediate register Pair D & E	0	0	0	1	0	0	0	1	10
LXI H	Load immediate register Pair H & L	0	0	1	0	0	0	0	1	10
STAX B	Store A indirect	0	0	0	0	0	0	1	0	7
STAX D	Store A indirect	0	0	0	1	0	0	1	0	7
LDAX B	Load A indirect	0	0	0	0	1	0	1	0	7
LDAX D	Load A indirect	0	0	0	1	1	0	1	0	7
STA	Store A direct	0	0	1	1	0	0	1	0	13
LDA	Load A direct	0	0	1	1	1	0	1	0	13
SHLD	Store H & L direct	0	0	1	0	0	0	1	0	16
LHLD	Load H & L direct	0	0	1	0	1	0	1	0	16
XCHG	Exchange D & E, H & L Registers	1	1	1	0	1	0	1	1	4
STACK OPS										
PUSH B	Push register Pair B & C on stack	1	1	0	0	0	1	0	1	11
PUSH D	Push register Pair D & E on stack	1	1	0	1	0	1	0	1	11
PUSH H	Push register Pair H & L on stack	1	1	1	0	0	1	0	1	11
PUSH PSW	Push A and Flags on stack	1	1	1	1	0	1	0	1	11
POP B	Pop register Pair B & C off stack	1	1	0	0	0	0	0	1	10
POP D	Pop register Pair D & E off stack	1	1	0	1	0	0	0	1	10
POP H	Pop register Pair H & L off stack	1	1	1	0	0	0	0	1	10
POP PSW	Pop A and Flags off stack	1	1	1	1	0	0	0	1	10
XTHL	Exchange top of stack. H & L	1	1	1	0	0	0	1	1	18
SPHL	H & L to stack pointer	1	1	1	1	1	0	0	1	5
LXI SP	Load immediate stack pointer	0	0	1	1	0	0	0	1	10
INX SP	Increment stack pointer	0	0	1	1	0	0	1	1	5
DCX SP	Decrement stack pointer	0	0	1	1	1	0	1	1	5
JUMP										
JMP	Jump unconditional	1	1	0	0	0	0	1	1	10
JC	Jump on carry	1	1	0	1	1	0	1	0	10
JNC	Jump on no carry	1	1	0	1	0	0	1	0	10
JZ	Jump on zero	1	1	0	0	1	0	1	0	10
JNZ	Jump on no zero	1	1	0	0	0	0	1	0	10
JP	Jump on positive	1	1	1	1	0	0	1	0	10
JM	Jump on minus	1	1	1	1	1	0	1	0	10
JPE	Jump on parity even	1	1	1	0	1	0	1	0	10

Mnemonic	Description	D7	D6	D5	D4	D3	D2	D1	D0	Clock[2] Cycles
JPO	Jump on parity odd	1	1	1	0	0	0	1	0	10
PCHL	H & L to program counter	1	1	1	0	1	0	0	1	5
CALL										
CALL	Call unconditional	1	1	0	0	1	1	0	1	17
CC	Call on carry	1	1	0	1	1	1	0	0	11/17
CNC	Call on no carry	1	1	0	1	0	1	0	0	11/17
CZ	Call on zero	1	1	0	0	1	1	0	0	11/17
CNZ	Call on no zero	1	1	0	0	0	1	0	0	11/17
CP	Call on positive	1	1	1	1	0	1	0	0	11/17
CM	Call on minus	1	1	1	1	1	1	0	0	11/17
CPE	Call on parity even	1	1	1	0	1	1	0	0	11/17
CPO	Call on parity odd	1	1	1	0	0	1	0	0	11/17
RETURN										
RET	Return	1	1	0	0	1	0	0	1	10
RC	Return on carry	1	1	0	1	1	0	0	0	5/11
RNC	Return on no carry	1	1	0	1	0	0	0	0	5/11
RZ	Return on zero	1	1	0	0	1	0	0	0	5/11
RNZ	Return on no zero	1	1	0	0	0	0	0	0	5/11
RP	Return on positive	1	1	1	1	0	0	0	0	5/11
RM	Return on minus	1	1	1	1	1	0	0	0	5/11
RPE	Return on parity even	1	1	1	0	1	0	0	0	5/11
RPO	Return on parity odd	1	1	1	0	0	0	0	0	5/11
RESTART										
RST	Restart	1	1	A	A	A	1	1	1	11
INCREMENT AND DECREMENT										
INR r	Increment register	0	0	D	D	D	1	0	0	5
DCR r	Decrement register	0	0	D	D	D	1	0	1	5
INR M	Increment memory	0	0	1	1	0	1	0	0	10
DCR M	Decrement memory	0	0	1	1	0	1	0	1	10
INX B	Increment B & C registers	0	0	0	0	0	0	1	1	5
INX D	Increment D & E registers	0	0	0	1	0	0	1	1	5
INX H	Increment H & L registers	0	0	1	0	0	0	1	1	5
DCX B	Decrement B & C	0	0	0	0	1	0	1	1	5
DCX D	Decrement D & E	0	0	0	1	1	0	1	1	5
DCX H	Decrement H & L	0	0	1	0	1	0	1	1	5
ADD										
ADD r	Add register to A	1	0	0	0	0	S	S	S	4
ADC r	Add register to A with carry	1	0	0	0	1	S	S	S	4
ADD M	Add memory to A	1	0	0	0	0	1	1	0	7
ADC M	Add memory to A with carry	1	0	0	0	1	1	1	0	7
ADI	Add immediate to A	1	1	0	0	0	1	1	0	7
ACI	Add immediate to A with carry	1	1	0	0	1	1	1	0	7
DAD B	Add B & C to H & L	0	0	0	0	1	0	0	1	10
DAD D	Add D & E to H & L	0	0	0	1	1	0	0	1	10
DAD H	Add H & L to H & L	0	0	1	0	1	0	0	1	10
DAD SP	Add stack pointer to H & L	0	0	1	1	1	0	0	1	10

NOTES 1 DDD or SSS B 000 C 001 D 010 E 011 H 100 L 101 Memory 110 A 111
2. Two possible cycle times (6/12) indicate instruction cycles dependent on condition flags

FIGURE 8.24 8080 Instruction Set (Courtesy Intel)

8080 INSTRUCTION SET

Summary of Processor Instructions (Cont.)

Mnemonic	Description	D_7	D_6	D_5	D_4	D_3	D_2	D_1	D_0	Clock[2] Cycles
SUBTRACT										
SUB r	Subtract register from A	1	0	0	1	0	S	S	S	4
SBB r	Subtract register from A with borrow	1	0	0	1	1	S	S	S	4
SUB M	Subtract memory from A	1	0	0	1	0	1	1	0	7
SBB M	Subtract memory from A with borrow	1	0	0	1	1	1	1	0	7
SUI	Subtract immediate from A	1	1	0	1	0	1	1	0	7
SBI	Subtract immediate from A with borrow	1	1	0	1	1	1	1	0	7
LOGICAL										
ANA r	And register with A	1	0	1	0	0	S	S	S	4
XRA r	Exclusive Or register with A	1	0	1	0	1	S	S	S	4
ORA r	Or register with A	1	0	1	1	0	S	S	S	4
CMP r	Compare register with A	1	0	1	1	1	S	S	S	4
ANA M	And memory with A	1	0	1	0	0	1	1	0	7
XRA M	Exclusive Or memory with A	1	0	1	0	1	1	1	0	7
ORA M	Or memory with A	1	0	1	1	0	1	1	0	7
CMP M	Compare memory with A	1	0	1	1	1	1	1	0	7
ANI	And immediate with A	1	1	1	0	0	1	1	0	7
XRI	Exclusive Or immediate with A	1	1	1	0	1	1	1	0	7
ORI	Or immediate with A	1	1	1	1	0	1	1	0	7
CPI	Compare immediate with A	1	1	1	1	1	1	1	0	7
ROTATE										
RLC	Rotate A left	0	0	0	0	0	1	1	1	4
RRC	Rotate A right	0	0	0	0	1	1	1	1	4
RAL	Rotate A left through carry	0	0	0	1	0	1	1	1	4
RAR	Rotate A right through carry	0	0	0	1	1	1	1	1	4
SPECIALS										
CMA	Complement A	0	0	1	0	1	1	1	1	4
STC	Set carry	0	0	1	1	0	1	1	1	4
CMC	Complement carry	0	0	1	1	1	1	1	1	4
DAA	Decimal adjust A	0	0	1	0	0	1	1	1	4
INPUT/OUTPUT										
IN	Input	1	1	0	1	1	0	1	1	10
OUT	Output	1	1	0	1	0	0	1	1	10
CONTROL										
EI	Enable Interrupts	1	1	1	1	1	0	1	1	4
DI	Disable Interrupt	1	1	1	1	0	0	1	1	4
NOP	No-operation	0	0	0	0	0	0	0	0	4
HLT	Halt	0	1	1	1	0	1	1	0	7

FIGURE 8.24 8080 Instruction Set (*cont.*)

194

TEST-FLOW CHART	FUNCTIONAL TEST DESCRIPTION	INSTRUCTIONS USED	
RESET	Reset microprocessor by stimulating the reset input.	None	
PROGRAM-COUNTER TEST	Increment the program counter through its full range.	NOP	
MEMORY TEST	Load directly the H and L memories with a 0. Then transfer both the H and L memories to all other memories. After each transfer, transfer that memory back to the H or L memory then output the H and L memories through the program counter. Increment H and L values by 1, reload and repeat test. Continue doing this until all 256 numerical combinations have been loaded into the H or L memory.	LX1 H PCHL MOVr1r2	
STACK-POINTER TEST	Transfer the H and L memories to the stack pointer and then verify that the stack pointer can increment and decrement through its full range. Verify by transferring through the H and L memories to the program counter.	LX1 H SPHL INXSP DCXSP PCHL DADSP	
ALU	Verify that the ALU will add, subtract, detect a 0, detect a positive value detect a negative value, perform carry, and perform all logical instructions.	ADDr ADCr SUBr SBBr ANAr XRAr ORAr	CMPr JC JNC JZ JNZ JP JM
ACCUMULATOR	Verify load, readback, rotate and transfer operations with all pattern combinations.	STAX B STAX D LDAX B LDAX D CMA	RLC RRC RAL RAR
TIMING AND CONTROL	Exercise all external stimuli and verify their correct action.	None	
INSTRUCTION DECODER	Performs all instructions that have not been previously exercised.	All others	

FIGURE 8.25 Test Flow-Chart (Architectural Test)

registers. To perform a complete functional checkout for all operations with all data patterns and between all register pairs, is out of the question. The skill is to select a representative set of operations to demonstrate:

(i) that the ALU can perform all its functions over a range of operand values, and

(ii) that transfer between the ALU and all source/destination registers can be made.

TIME AND CONTROL. The tests here can include not only timing measurements on the response to external stimuli but also measurement of internal timing values. In common with many other microprocessors, the amount of time taken to execute an 8080 instruction varies from instruction to instruction. Some instructions, such as NOP, are executed in a minimum number of machine cycles whereas others, such as register–register exchange (e.g., XTHL) require the maximum number. Tests can be included to verify that each instruction is at least completed within the specified number of machine cycles.

INSTRUCTION DECODER. Most instructions will have been used in earlier tests. Those that have not are tested here.

8.7.2 Instruction Code Checkout

An alternative method for approaching the construction of a checkout sequence for a microprocessor device is to demonstrate that each instruction is executed correctly for a representative data set. In essence, this was one of the objectives of the previous strategy, but program structure was obtained by considering the architectural features of the device rather than its operational features.

The problem with an instructional checkout technique is: where do you start? (To a certain extent, this problem exists with the architectural strategy.) One solution to the ordering problem is to rank each instruction in terms of those sections of the chip which must be working in order for the instruction to work. In essence therefore, the least-demanding instruction is checked first and, if this works, it can be used to test more-demanding instructions, and so on until the complete instruction set is checked. A technique such as this is known as a "bootstrap" technique. It requires some scoring mechanism to allocate a score to each instruction and hence determine the bootstrap order. One possible scoring scheme is as follows: initially, all instructions have zero score.

(i) For each byte required by the instruction and its operands, add 1 point

(ii) For each register (excluding the instruction register and program counter) affected by the instruction, add 1 point

(iii) For each memory access in addition to the initial read requirement, add 1 point

(iv) For each clock cycle required, add 1 point

(v) For each conditional flag that could be affected, add 1 point (up to a maximum of 2)

(vi) For each microinstruction control line required, add 1 point

(vii) For each additional instruction register or program counter access (over and above the standard requirements), add 1 point

Figure 8.26 shows the result of applying this scoring technique to the 8080 instruction set. (See Figure 8.24.) This score is used to produce the testing sequence shown in Figure 8.27. Instructions with the lowest score are tested first. Others follow according to the score determined in Figure 8.26.

8.8 Receiver/Transmitter Devices (UARTs)

The UART consists of two sections—transmitter and receiver—and is shown in Figure 8.28.

The transmitter section converts parallel input data into serial output data and formats the data with a single start bit (0), data bits, even or odd parity bit, and one or two stop bits (1s). The receiver section accepts serial data and converts it into parallel form, again with appropriate start, stop, and parity bits. Most UARTs are able to receive and transmit simultaneously if required (full duplex operation). To test a UART, the following tests, at least, should be performed.

Test 1 Transmitter Section

Read in and clock out: all-0s, all-1s, chequerboard (0101 . . .), inverse chequerboard (1010 . . .) with variations on the length of the word, number of stop bits (one or two), and parity bit (odd or even). During the test, monitor the status of the various flags. The tests should be carried out at various speeds up to the maximum baud rate, and with arbitrary gaps between the input data to simulate true asynchronous input.

SCORE	INSTRUCTIONS
5	NOP
6	CMA; CMC; DI; EI; STC
7	DCX SP; INX SP; RAL; RAR; RLC; RRC
8	CMP r; DAA; DCX (B,D,H); HLT; INX (B,D,H); MOV r_1, r_2
9	ADV r; ADD r; ANA r; DCR r; INR r; ORA r; PCHL; SBB r; SPHL; SUB r; XCHG; XRA r
10	MVI r;
11	CPI
12	ACI; ADI; ANI; MOV r,M; ORI; SBI; STAX: SUI; XRI
13	CMP M; LDAX; MOV M,r; RST
14	ADC M; ADD M; ANA M; DAD H; IN; JMP; LXI SP; ORA M; OUT; RET; SBB M; SUB M; XRA M
15	DAD SP; JC; JM; JNC; JNZ; JP; JPE; JZ; LXI(B,D,H),; POP
16	DAD(B,D); MVI M,
17	DCR M; INR M; PUSH
18	LDA; STA
21	LHLD
23	SHLD; XTHL
24	CALL
10/16	RC; RM; RNC; RNZ;) RP; RPE; RPO; RE)See
19/25	CC; CM; CNC; CNZ;)Note. CP; CPE; CPO; CZ)

Note: Conditional CALLS and RETURNS have two instruction cycle times, depending on whether or not the appropriate flag is set.

FIGURE 8.26 Scores for 8080 Instruction Set

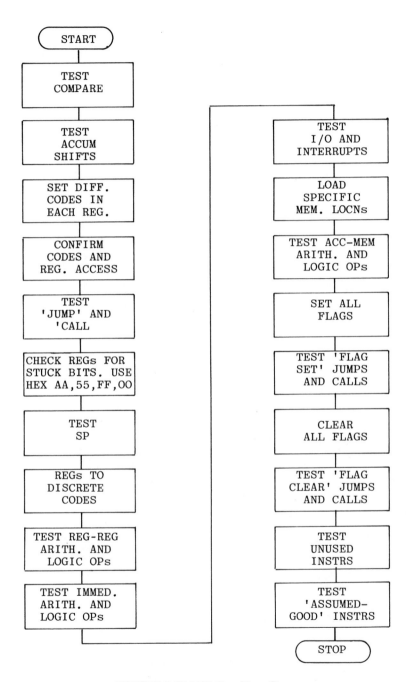

FIGURE 8.27 8080 Test Flow-Chart

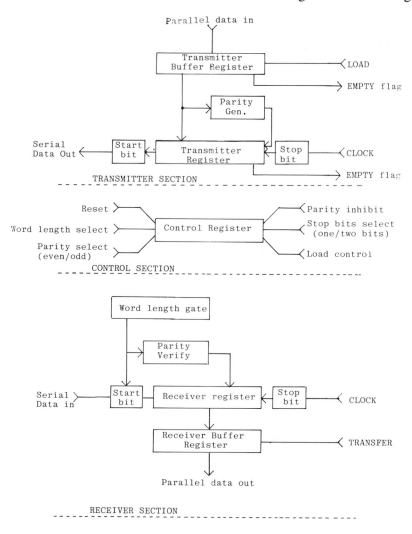

FIGURE 8.28 UART Organization

Test 2 Receiver Section

Similar to the transmitter section. A particular problem with this section is to test the action of the error flags. For example, a framing error is flagged if a 0 instead of a 1 is encountered in the stop bit position. The serial input data should include at least one example of this error to test the action of the framing error flag logic. A similar strategy can be used to test other error flags such as incorrect parity and overrun.

8.9 Controller Devices

Controller devices exist in many forms and for a variety of applications. For example, special-purpose controllers exist for direct memory access (DMA), cathode-ray tubes (CRT), hard and floppy disc, IEEE 488 bus, priority asynchronous interrupts, etc. There is little point in considering each of these controllers in any detail in this book. The general test procedure follows the previous processes: understand what the device is supposed to do and how it does it and write an appropriate set of test patterns.

One of the main difficulties with controller devices is that they often interface between sub-systems operating at different speeds. To be realistic therefore, the device should be tested at the expected operating speed. In effect, the tester has to be programmed to "look like" a floppy-disc transport, say. This means that the test programmer not only needs to understand how the controller device works, but also what the requirements of the floppy-disc transport might be. In this respect therefore, the test programmer needs to be a very competent digital systems engineer.

8.10 A General-Purpose Test Strategy

Finally, we present a general-purpose test strategy for a bus-structured board containing LSI/VLSI devices. In particular we will discuss an ordering for the overall test strategy which will allow an orderly progression of tests. The strategy will be illustrated with reference to the Intel 80/20 Single Board Computer based on the 8080A microprocessor device. The general structure of this board is shown in Figure 8.29.

The main steps in the test strategy are as follows:

STEP 1 Isolate all devices (cpu and non-cpu) from the system busses and test the bus for stuck-at and bridging faults.

The busses are the main veins and arteries of the board. Therefore,

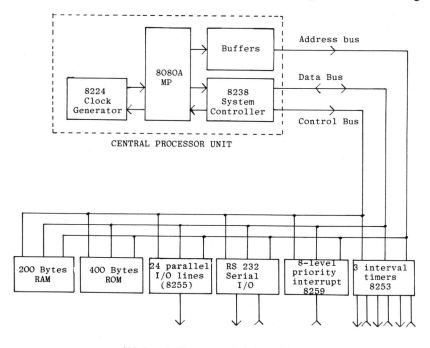

All bus buffers are tristateable.

FIGURE 8.29 Intel 80/20 Single Board Computer

if there are faults on the bus it will not be possible to control and observe individual devices.

Bus isolation is achieved by placing each bus buffer into its high-impedance (tristate) mode, or even by removing some of the devices. Care should also be taken to ensure that as few faults as possible on other chips can cause a chip-select node to become asserted inadvertently. Access to the bus can either be direct (if the busses are brought out to edge-connector positions) or indirect by either a multi-pin clip over one of the bus devices or a multi-pin plug that plugs into a socket vacated by one of the devices.

The test for the bus consists of a test to show that each line can be pulled both high and low. This requires either that the tester has appropriate pull-up and pull-down facilities or that a special test fixture is available.

Failure to pass this test usually means one of two things: either the bus contains a genuine stuck-at or node-to-node bridging fault, or one (or more) of the devices that is supposedly in a tristate mode is not. Even at this early stage, therefore, diagnosis of a fault can prove troublesome and requires the use of programmed suggestions as to the possible cause of the fault, as well as the use of current-tracing probes.

STEP 2 Maintaining the cpu in its tristate mode, test each non-cpu device by means of an appropriate low-speed or high-speed test.

The objective is to verify the behavior of each non-cpu device before verifying the behavior of the microprocessor.

This works fine provided the tester has good access to each device and that all devices except that being tested can be held in their isolated mode. The order of testing the devices is to test the simplest first and gradually build up to the most complex. The aim should be to complete all unidirectional tests (data flow in one direction only) before commencing the bidirectional tests. This has the advantage of finding or eliminating simpler types of fault, such as nodal stuck-at and bridging faults, before embarking on the tests for more complex faults, such as internal failure of a device. Probing sequences are also shorter for unidirectional tests than for bidirectional tests. For the 80/20 board, an ordering could be: serial I/O, parallel I/O, timers, I/O expansion, priority interrupt controller, ROM, RAM.

As each device is tested and passed, it can be used to assist in the testing of subsequent devices. In this way, interaction between devices can be tested, thereby allowing possible exposure of certain timing faults. Note that it is vital that the tester has direct and total control over the control bus.

STEP 3 Tristate all non-cpu devices, reinstate and initialize the cpu, and carry out a cpu test.

Again, the test may have to be limited simply because a full test will require other non-cpu devices, such as the priority interrupt controller. Nevertheless, a certain amount of cpu testing can be carried out to gain some confidence in the correctness of the set of devices. In general, the tests at this stage are initially at a low frequency, i.e., single-step where feasible,* or at the lower end of the operating frequency range. It

*Not all microprocessors have a single-step capability—the Motorola 6800, for example, has not.

may also be possible to apply some real-time tests, in which case internal cpu timing problems may be detected.

Note, however, that even initialization may not be as simple as expected. For example, if the reset line of the Rockwell R6502 micro-processor is pulled low (to reset) and then returned high, ''the internal states of the machine are unknown and all registers, except the program counter, must be re-initialized.'' Part of the reset action is to load the program counter with a fixed address (FFFC in PCL, FFFD in PCH) and, as the reset line goes high, this address points to the start of the user code in memory. Execution of this code starts after a delay of 6 machine cycles.

STEP 4 Reinstate all the devices and carry out a full system test at operational speed.

By the end of Step 3, we have some confidence that each major device on the board can operate correctly when looked at in isolation from all other devices. The final stage therefore is to load a test program into the onboard RAM and carry out program execution at real-time speed. At this level, the program is really testing for real-time inter-action problems and is aiming to complete any device functional tests that were incomplete at Steps 2 or 3. Within Step 4 therefore, the progression of tests should be structured to work from easiest to most difficult. In this context, ''easiest'' means the test involving the smallest number of interacting devices.

As with other steps, the most difficult problem is to identify the precise cause of failure; the test programmer has to employ all the diagnostic aids available—programmed message, automatic guided probe, current-sensing probes, and possibly oscilloscopes and state-analysers. The latter are useful to examine the performance as timing and voltage supply lines are taken to their design margins. Tests such as these are often very good at identifying interactive timing problems or marginal components.

8.11 Final Comments on Testing Bus-Structured Boards

Techniques for testing bus-structured boards containing LSI/VLSI de-vices are still largely manual and require a high level of understanding and test-programming skill on the part of the test programmer. In addition, the complex interaction between devices operating in the MHz region creates many debugging problems when generating the test. Even when the test program is thought to be complete, there is no way to evaluate how it performs. Simulators are not yet able to model all possible fault conditions and physical fault insertion is limited to very basic faults. Precise figures for fault cover are therefore unobtainable. It is also very difficult to evaluate the diagnostic performance of the test program, again because of the limitations of logic simulators and the restriction on the range of physical faults. Finally, the whole process is periodically thrown into a state of confusion by the announcement of some new form of internal-device failure. Without doubt, therefore, the processes of verifying correctness of operation will remain a challenge for some years.

Chapter 9
Design of Testable Logic

In this chapter, we will look at how many of the problems of testing can be eased or even removed by due consideration of the stages leading up to and including the design of a printed-circuit board. For convenience, we will identify two main phases in the design process. The first is the conversion of an abstract logical requirement into a logic diagram; the second, the realization of the logic diagram as a PCB. These two phases will be referred to as the "circuit design" phase and the "board design" phase respectively and we will discuss how design for testability can be incorporated into either or both of these phases.

9.1 What is Testability?

There are many definitions of testability—some formal, others informal. Essentially, however, testability relates to cost—the cost of generating test patterns to meet certain criteria; the cost of ensuring correct diagnosis; the cost of providing support tools; etc. An informal definition of testability is as follows.

A logic circuit is "testable" if a set of test patterns that guarantee detection, and unambiguous correct location if required, of a predefined set of fault conditions can be generated, evaluated, and used in a cost-effective manner.

There are many points of clarification required to make this definition workable and different companies or application environments will emphasize different areas. What is cost-effective for one company may be completely unacceptable to another. Nevertheless, the central point remains, namely, that the various costs associated with testing contribute to the final testability of the circuit. Any procedure for lowering these costs can therefore be construed as a "design for testability" constraint.

A more formal definition of testability is emerging in the literature and is based on an assessment of the "controllability" and "observability" features of a circuit design. (See Figure 3.17, Chapter 3.) Controllability is defined as the ease, or otherwise, of setting a particular internal logic node to either logic 1 or logic 0. Observability is defined to be the ease, or otherwise, of observing the response of an internal logic node. An internal node is controlled from the primary inputs and observed at the primary outputs and the process of test-pattern generation relies on the ability both to control and to observe each node in the circuit. A measure for nodal testability can therefore be quantified in terms of the nodal controllability and nodal observability values. Circuit testability can then be determined from a knowledge of the testability of all nodes in the circuit.

In essence therefore, testability measures based on controllability and observability features are really only a measure of the ease or otherwise of generating test patterns, rather than of evaluating or using the patterns. Nevertheless, there would seem to be many uses for such measures, such as:

(i) advising on the better of two designs (a revised design may have a higher testability rating than the original);

(ii) allowing a judicious selection of test points (nodes with low observability are obviously good candidates);

(iii) identifying potentially difficult nodes to test (low controllability and observability).

The advanced nature of testability measures based on the evaluation of controllability and observability precludes any detailed description in this book. A general comment, however, is that, to be useful, a measure should be inexpensive to compute in comparison with the costs of deriving the tests. In this way, the measure can be used as an interactive design aid.

In the remainder of this chapter we will look at procedures and guidelines for designing testable logic circuits.

9.2 Designing Testable Circuits

There have been many suggestions in the literature for designing logic circuits that are testable, or even "easily-testable." Many of these proposals have related only to combinational circuits or have incurred too high a penalty in terms of extra primary inputs, extra primary outputs, or extra gates. In practical designs, there are often very few spare edge-connector positions or integrated-circuit positions, and any increase in the use of either of these commodities to increase testability must be balanced against the corresponding decrease in reliability and increase in production costs. Because of this, most of the suggestions have not been adopted by industry as practical techniques. Two that have, however, are "scan-in, scan-out" (SISO) and "signature analysis" (SA). These are discussed in the following sub-sections.

9.2.1 Scan-In, Scan-Out (SISO) Design

SISO is a deliberate attempt to reduce the problems of generating test patterns for logic circuits containing stored-state devices and global feedback. The philosophy of the technique is one of "divide and conquer" and will be described with reference to the general model for the clocked (synchronous) logic circuit shown in Figure 9.1.

In this diagram, the major elements of the circuit have been identified as two blocks of combinational logic together with a third block containing the stored-state devices. The outputs of the stored-state devices are referred to as "secondary variables" and become inputs to both the output logic block and the next-state logic block. In this way, both the future value of the secondary variables and the present value of the primary outputs are a function not only of the present value of the primary inputs but also of the present value of the secondary variables.

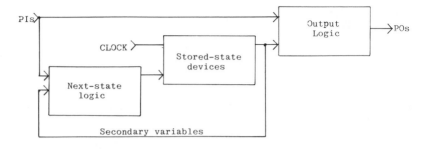

FIGURE 9.1 General Model for a Logic Circuit

Every synchronous logic circuit, even the most simple form of combinational circuit, can be seen as a variation of the model shown in Figure 9.1. The variations are formed by omitting various blocks of logic or various connections. For example, Figure 9.2 shows how the self-starting divide-by-5 counter circuit used in Chapter 3 (Example 6) could be drawn to conform to the model.

Note that there are no primary inputs (other than the CLOCK) and that the combinational logic producing the primary outputs is effectively a set of short-circuit connections.

Returning to Figure 9.1, we can now see what the problems are with regard to test-pattern generation. The only inputs over which we have direct control are the primary inputs. Similarly, the only outputs we are able to observe directly are the primary outputs. Unfortunately, in the general case, both blocks of combinational logic are controlled by the secondary variables as well as by the primary inputs and, further-more, the values of the secondary variables are determined by the next-state logic. In other words, we have a ''chicken and egg'' problem. In order to test the stored-state devices, we have to be able to set them into a known state: in order to test the next-state logic (and the output logic) we need control of the values of the secondary variables. The question is—which do we do first; also, what happens if we discover certain secondary variable values to be impossible to generate? (For example, the very nature of the divide-by-5 counter does not allow the values 101, 011, or 111 on A, B, and C. These are the illegal start states.)

The SISO technique provides a solution to these problems by

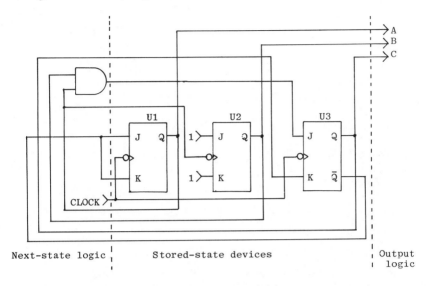

FIGURE 9.2 Redrawn Version of Divide-by-5 Counter

reducing the complexity of the circuit structure. The principle of the technique is to provide additional facilities in the circuit such that:

(i) the stored-state devices can be tested in isolation from the rest of the circuit;

(ii) the future values of the secondary variables can be set to any desired level independent of their present value;

(iii) the response of the next-state logic can be observed more directly than through the primary outputs.

The method by which this is achieved is shown in Figure 9.3.

Leaving aside the detail for the moment, we can see that facilities have been added to isolate the stored-state devices from the rest of the circuit and, at the same time, to reconfigure these devices into a synchronous shift register. In its simplest form, the shift register is serial-in, serial-out, and three additional primary access points have been provided. The first, called SCAN-IN, allows data to be loaded into the serial shift register, whereas the second, called SCAN-OUT, allows observation of the contents of the shift register. The other input is a

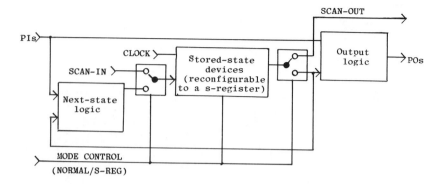

FIGURE 9.3 Principle of Scan-In, Scan-Out

mode control input which establishes the NORMAL or shift register
(S-REG) configuration of the circuit. The switchover from one mode to
another is implemented with 2-way multiplexers and decoders, shown
figuratively as ganged switches in Figure 9.3. The testing strategy is
now as follows:

STEP 1   Switch to S-REG mode. Test the operation of all stored-state
          devices by standard shift-register tests (e.g., clock 0's, 1's,
          chequerboard).

STEP 2   Determine a set of tests for the next-state logic, assuming:
          (i) total control of all inputs (primary and secondary),
          (ii) direct observation of outputs.
          Apply each test in the following manner:

STEP 2A   S-REG mode. Preload secondary variables to the required
          values.

STEP 2B   Establish correct values on primary inputs.

STEP 2C   Switch to NORMAL mode. Next-state logic will now re-
          spond to primary and secondary input values. Clock these
          values into the stored-state devices.

STEP 2D   Switch back to S-REG mode and clock out the contents of the
          stored-state devices. Compare with the expected fault-free
          values.

STEP 2E Repeat Steps 2A–2D for next test until all tests have been applied.

STEP 3 Repeat Step 2, this time with tests for the output logic. Note here that observation of the response is direct at the primary output rather than indirect via the stored-state devices.

We can now see the "divide-and-conquer" philosophy more clearly. Rather than test the circuit as one conglomerate of devices connected in a complex fashion, SISO allows each segment of the circuit to be tested separately and in a methodical way. Furthermore, if we assume a standard test for the stored-state devices, the only circuits that require test-pattern generation are combinational ones. This means that algorithms, such as the D-algorithm, can be used to provide the tests. All the test programmer need do is to embed the results into the general strategy just described.

Before commenting on the possible disadvantages of SISO, we will illustrate in more detail how the technique could be implemented. Figure 9.4 shows a conventional design for a synchronous 3-bit twisted-ring counter based on D-type flip-flops.

Figure 9.5 shows how this circuit could be modified to suit a SISO test strategy—in this case, very simply by including a 2-way multiplexer as shown.

The following test strategy can now be employed to test the circuit:

Step 1 S-Reg mode. Initialize ABC to 000
        Change sequence $000 \to 100 \to 110 \to 111$ and check on A,B,C

Step 2 Apply individual tests to next-state logic.
        A suitable set of tests is given by

Test No.	A B C	NS
1	1 1 0	1
2	0 1 0	0
3	1 1 1	0
4	0 0 0	1
5	0 0 1	0

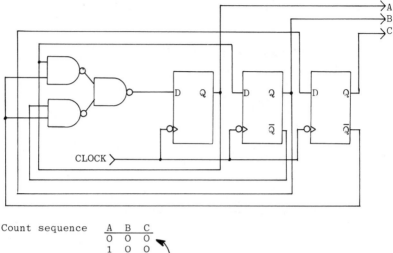

Count sequence

A	B	C
0	0	0
1	0	0
1	1	0
1	1	1
0	1	1
0	0	1
0	1	0
1	0	1

Illegal states

FIGURE 9.4 Synchronous 3-Bit Twisted-Ring Counter

This set covers all single stuck-at faults in the next-state logic.

Step 3 Test 2-way multiplexer by setting NS and SI to opposite values and clocking output into first D-type.

The example appears to be relatively trivial but, in fact, it is not. At first sight, it would seem easier to check the count sequence rather than to modify the circuit and employ the strategy shown above. If this were done however, certain faults in the next-stage logic would go undetected. The faults are connection x s-a-1 or connection y s-a-1, or both. (The s-a-1 faults could be the result of a track open-circuit.) The only way these two faults can be detected is to set the three flip-flops into one of the illegal states, namely A = 0, B = 1, C = 0. Without the SISO modification, this is not possible; with the modification, however, it is (Test 2 above).

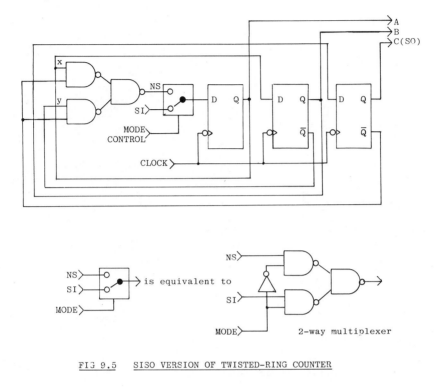

Another observation we can make about this example is that the modification to SISO form was made easy because the flip-flops were already configured as a shift register. This is not always the case, as can be seen in the circuit for the divide-by-5 counter (See Figure 9.2). The clock for the middle JK flip-flop, U2, comes from the Q output of U1 and not from the system clock. Also, the general configuration of the flip-flops is not that of a simple shift register. Modifying this circuit to a SISO form is therefore more complex and requires the use of both multiplexers and decoders, as shown in Figure 9.6.

For this example, it could be argued that the SISO modification does not enhance the testability of the circuit. In its original form, a straight checkout of the count sequence will also test for all stuck-at faults on the 2-input AND gate. The SISO version offers no significant advantage therefore: indeed, it is positively disadvantageous from a reliability and parts-count point of view.

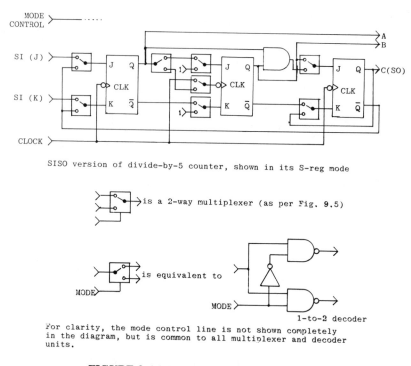

SISO version of divide-by-5 counter, shown in its S-reg mode

is a 2-way multiplexer (as per Fig. 9.5)

is equivalent to

MODE

MODE

1-to-2 decoder

For clarity, the mode control line is not shown completely
in the diagram, but is common to all multiplexer and decoder
units.

FIGURE 9.6 SISO Version of Divide-by-5 Counter

In summary therefore, SISO may offer significant test advantages
if the circuit already contains a shift register form of structure. However,
the penalties for introducing SISO features into a circuit design are:

(i) additional primary input/output requirements;

(ii) a possible increase in the length of time required to test the
circuit (although this can be speeded up by designing the circuit
so as to allow parallel load and parallel read-out in the S-REG
mode);

(iii) additional logic devices such as multiplexers and, possibly,
decoders;

(iv) greater parts-count and connect-count, leading to a lower overall reliability figure.

Despite these penalties, SISO is a powerful technique for designing highly sequential logic that is testable, provided the design is

constrained at the outset. A number of companies have developed the design process to a high level of sophistication. Perhaps the most advanced technique is the Level-Sensitive Scan Design process developed by IBM, the general principles of which are discussed in the following section.

9.2.2 Level-Sensitive Scan Design (LSSD)

There are in fact two fundamental constraints to the LSSD technique. The first is that all changes in the state of the circuit take place according to the level of a clock control signal, rather than to an edge change. Furthermore, the steady-state response to an input change is independent of any propagation delays (gate and interconnect) in the circuit, and is also independent of the order of input-value changes in the event of simultaneous multiple changes. This is the property of "level-sensitivity" and it is designed to reduce the dependency of the circuit on its ac parameters, such as degraded rise and fall times, degraded propagation delays, or other faults that introduce race or hazard conditions. In this way, the effect of timing faults on the operation of the circuit is reduced.

The second constraint is that the circuit should possess SISO characteristics for test purposes (called the "scan" property) and, in this respect, a special-purpose stored-state device has been designed and is the only stored-state device allowed. This special-purpose device is called a polarity-hold Shift-Register Latch (SRL) and is shown in block diagram form in Figure 9.7, and in NAND-gate equivalent in Figure 9.8.

The SRL consists of two D-type latches, L1 and L2. L1 constitutes the normal stored-state holding device with system data and system clock inputs and Q data output. The latch operates normally provided shift clock A is 0. Shift clock B is also held at 0 during normal operation.

To operate the latch as part of a shift-register chain, shift clock A is set to 1. This enables Scan Data In information to be latched directly into L1. Shift clock A is then changed back to 0 (to hold the value in L1) and shift clock B raised to 1. The contents of L1 are then latched into L2 and will be held permanently as shift clock B falls back to 0.

In a practical circuit, the SRLs are connected permanently to form

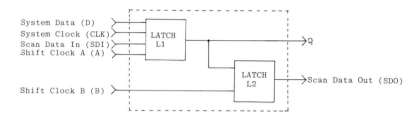

FIGURE 9.7 Polarity-Hold Shift-Register Latch (SRL)

the shift register by connecting the Scan Data Out output of one SRL to the Scan Data In input of another. The two shift clocks, A and B, are common to all SRLs. The general format of the circuit is shown in Figure 9.9.

In practice, there are many design rules associated with the implementation of LSSD. To be effective, LSSD must be imposed as a rigid design discipline, supported by a design automation system to check for violations of the design rules. Also, there are several variations on the basic scheme that offer advantages with regard to sensitivity to timing faults. In particular, the design can be partitioned into two

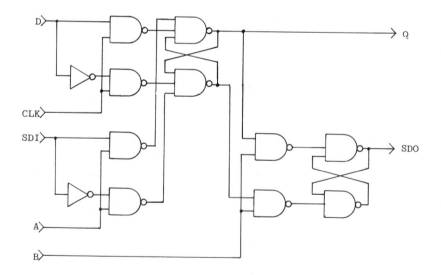

FIGURE 9.8 Nand Equivalent of SRL

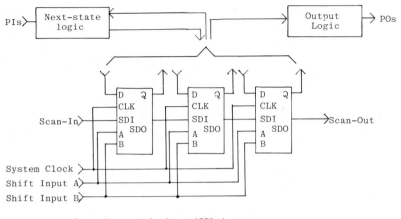

Stored-state devices (SRLs)

FIGURE 9.9 General Structure of an LSSD Circuit

distinct sections, each of which is controlled by one phase of a two-phase clock system. The general structure of this particular form of implementation is shown in Figure 9.10. Note that there is no direct feedback from S_1 to C_1 or from S_2 to C_2.

9.2.3 Final Comments on LSSD

Without doubt, implementing a SISO design policy, such as LSSD, imposes penalties of extra pinouts on each chip together with extra gates

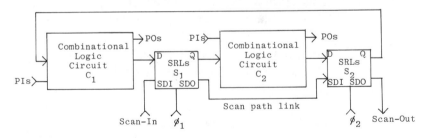

Combinational logic circuits C_1 and C_2 contain both next-state and output logic. Shift inputs A and B are not shown but are common to S_1 and S_2

FIGURE 9.10 Two-Phase Clock Implementation of LSSD

within each SRL device. The technique also places constraints on the freedom of choice traditionally associated with logic design. Indeed, the cynics would argue that LSSD has removed much of the creativity of logic design. That this is not true can be seen by consideration of the following questions, most of which have not yet been answered.

1. Do synthesis algorithms exist to convert a logic specification written in some high-level register-transfer language into an LSSD design and, if so, what is the best form of implementation?

2. If alternative algorithms exist to carry out a particular data processing requirement (such as restoring and non-restoring binary division algorithms), which one is better suited for an LSSD implementation?

3. Following from 2, are there any general characteristics for data-processing algorithms that lend themselves to an LSSD form of implementation?

4. In its simplest form, the L2 latch of an SRL performs no useful function except in the S-REG mode of operation. Is it possible to make some operational use of the latch, either as it exists or with the addition of further logic within the latch?*

5. What are the optimum designs for standard sequential circuits, such as the many varieties of counter, and how can these standard designs be embedded in a general synthesis algorithm?

In summary, therefore, techniques such as LSSD still present a number of challenges to logic designers and yet offer significant advantages both in the way in which the circuit operates and in the testing. In addition, logic simulation for both design verification and fault-cover evaluation is straightforward, and field diagnosis is simplified. Against advantages of this kind, the penalties do not seem too onerous.

9.2.4 Signature Analysis

Signature analysis (SA) is a guided probe technique, based on the use of CRC signatures, for isolating faults that have developed in operational

---

*Recent publications on LSSD have suggested that this is possible. As it stands, the L2 latch can be used either as a backup buffer register or, with the addition of a 2-input OR gate and second system clock, as a master-slave storage element.

boards. In essence therefore, it is a troubleshooting technique for field maintenance and is based on the same process as the signature-testing technique described in Chapter 5. The main difference is that the test stimulus is supplied by devices on the board (rather than from a tester) and the support equipment required is a system for capturing and displaying CRC signatures. Such an instrument is simple to design and build and is certainly portable. Figure 9.11 shows the main features of a circuit designed to SA standards.

There are two additional design requirements in Figure 9.11. The first is the provision of the onboard test stimuli as a Built-In Test. These can be either the outputs of some form of a binary counter or the outputs from a ROM. These outputs are gated into the main control and data inputs to the circuit.

The second addition is a facility to break all the feedback loops in the circuit (by switches, jumpers, or tristate buffers). This is done to avoid the loop-breaking problem discussed in Chapter 5.

The fault-isolation strategy is now quite simple. The service engineer places a suspect board into its test mode of operation. In this mode, the test stimulus is applied to the rest of the circuit and each node will respond accordingly with a sequence of 0s and 1s. Working from an

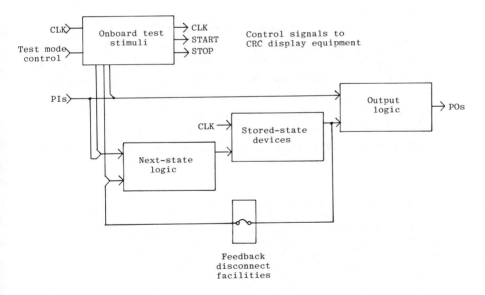

FIGURE 9.11 Circuit Designed for Signature Analysis

annotated circuit diagram or from a troubleshooting tree, the service engineer probes back through the circuit looking for the device whose input signatures compare correctly with the reference values but whose output signature does not. In this way, the source of the fault is located.

In its simplest form, the test stimuli do not need to establish sensitive paths through the circuit. Provided each node changes its value at least once during the application of the stimuli, s-a-0 and s-a-1 faults can be located by probing every node and manually resolving cause-effect relationships. If, in addition, the stimuli create sensitive paths through to the primary outputs, then the service engineer can follow the same probing algorithm as described in Chapter 5 and produce an accurate diagnosis without necessarily probing every node.

The CRC display equipment will need three control signals from the board—the system CLOCK, a START signal, and a STOP signal. The START and STOP signals determine the time window during which data should be strobed into the CRC shift-register contained within the instrument. There should also be some provision to handle nodes that go to a high-impedance tristate level sometime during the START-STOP window.*

Figures 9.12 and 9.13 show two of the new range of portable instruments designed to support signature analysis as a diagnostic technique.

As with SISO, the decision to include SA into a design should be taken before implementation; it is difficult to implement SA facilities retrospectively. Unlike SISO, SA is not intended to simplify the problems of test-pattern generation. It is a technique to simplify field diagnosis of faults and, in this respect, has the advantage of low-cost equipment and requires relatively low skill. The penalties are increased design time and hardware costs but, as with SISO, these penalties are relatively insignificant for certain types of digital equipment.

---

*The Hewlett-Packard instrument (HP5004A Signature Analyzer) achieves this by making the guided probe look like a 50KΩ impedance pulled to a nominal 1.4v reference value. The probe contains a dual-threshold comparator and, when the node changes from an active state to a floating high-impedance state, the comparator still considers the input to be the same as the last valid active state, i.e., a valid 0 followed by a floating state is considered to be a sequence of valid 0s, whereas a valid 1 followed by a floating state is considered to be a sequence of valid 1s.

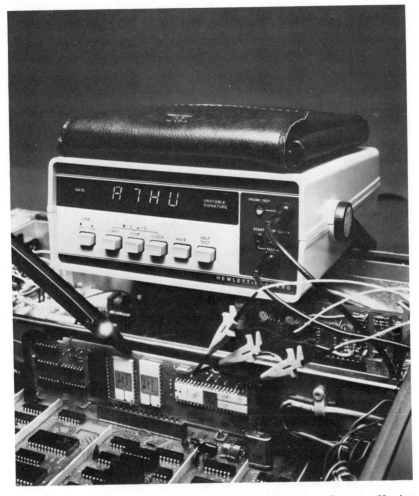

FIGURE 9.12 Hewlett-Packard 5004A Signature Analyzer (Courtesy Hewlett-Packard)

9.2.5 Application of SA to Bus-Structured Boards

Signature analysis has particular application to bus-structured boards, such as memory or microprocessor boards, because of the inherent ease of applying and monitoring the tests. Also, the tests can be preloaded

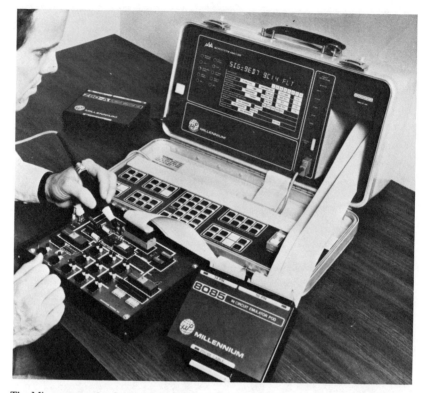

The Microsystem Analyzer—Series 4000 from Millennium Systems, Inc. is a universal microprocessor test instrument combining the three basic approaches to microsystem trouble-shooting—time domain analysis, in-circuit emulation, and signature analysis— in a single portable instrument.

FIGURE 9.13 Millennium μSA Microsystem Analyzer (Courtesy Microsystem Services

into existing RAM. Alternatively, the tests can be resident in part of an on-board ROM (or the user ROM can be replaced by a test ROM). If the tests are resident, then effectively the board can become self-testing if some means of failure indication, such as an LED, is provided.

The order of testing (as noted in the previous chapter) is obviously important and for self-test routines requires at least the ''kernel'' of the board to be working. This is the minimum configuration necessary to start the test program. If the kernel is working then it can be used to test other devices.

It is possible to take self-testing a stage further and incorporate self-test facilities directly onto a chip. An example of this occurs with the Motorola MC6805 microcomputer chip. This device is derived from the MC6800 chip but it includes 64 bytes of RAM, 1100 bytes of user ROM, 20 I/O lines, an 8-bit timer, power-on reset, interrupt control, and 116 bytes of ROM holding the self-test program. The test mode is invoked by raising the voltage on the timer pin to 8v and applying a reset signal. The self-test program then runs continuously and tests the reset function, I/O ports, RAM (walking bit pattern), user ROM (odd-1 parity checksum), and interrupt facilities. If the device passes these tests, a 3Hz square wave is produced on one of the I/O lines. This signal can be used to drive an LED. Errors are identified by the coded output of two of the I/O lines. The codes correspond to ROM error, RAM error, I/O error, and interrupt error.

It is claimed that the self-test program covers every addressing mode and nearly every instruction of the cpu. In chip size, the self-test ROM occupies 1% of the total die size.

9.3 Designing Testable Boards—Practical Guidelines

In this section, we shall present a number of practical guidelines for improving the testability of a digitial PCB. Many of these techniques are simply engineering commonsense and have been hinted at in earlier chapters. It is useful to collect them together, however, into a single checklist. The guidelines are classified into two types; those that aid test-pattern generation and those that aid test application and fault finding. (There does not seem to be very much that can be done to assist test-pattern evaluation, except the obvious one of providing suitable simulation support tools.)

9.3.1 Aids to Test-Pattern Generation

Guideline 1. Maximize access to the inputs or outputs of internal stored-state devices and other more complex devices

We now know that the ability to generate tests is heavily dependent on the ease or otherwise of being able to control and observe stored-state devices as well as other more complex devices such as shift registers, counters, ROMs, RAMs, microprocessors, and associated devices. If there are spare edge-connector positions, then these should be utilized wherever possible to improve controllability and/or observability of these devices. If test points are to be used to increase observability, these should be located at strategic positions such as

    (i) major control lines (system clock, master reset),
   (ii) logically redundant nodes (see Guideline 2),
  (iii) flip-flop or counter outputs,
  (iv) the trunk of points of high fanout.

If edge-connector positions are limited, test-point data can be "funnelled" off the board either via a multiplexer or via a parallel-load, serial-out shift register. (See also Guideline 18.) Single-pin or multi-pin clips direct onto the leads of integrated-circuit devices can also be used to increase observability, but their use will add to the requirements of the interface between the board and the tester.

Note also that a test point can be used for control purposes as well as for observation purposes. A test point on a node between two standard TTL devices can be driven low (but not high) to allow partial control of the driven device. Effectively, we are making use of the technique for inserting physical faults discussed in Chapter 4.

Other important control points are as follows:

    (i) The ENABLE/HOLD line of a microprocessor.
   (ii) The control, address and data busses of a microprocessor board.
  (iii) The ENABLE and R/W lines of a RAM device.

Control of the RAM address lines is also useful since it allows the RAM to be used as a single-cell device. A collection of RAM chips therefore can be made to resemble a parallel-in, parallel-out register.

Consideration should also be given to the use of a second edge-connector, along another edge of the board, to provide additional access to or from the board. A disadvantage with this approach is that it increases the interface requirements between the board and the tester. (See Guideline 21.)

Guideline 2 Avoid logical redundancy

A circuit node is logically redundant if all the outputs of the circuit are independent of the binary value on the node for all input combinations or state sequences.

Logical redundancy often exists in circuits either intentionally, for example to mask a static-hazard condition, or unintentionally. The problem with a logically-redundant node is that, by definition, it is not possible to make an output value dependent on the value of the redundant node.

This means that certain fault conditions on the node cannot be detected, thus creating two problems. The first is that the fault condition may reintroduce the hazard condition it was designed to eliminate; the second, that the fault on the redundant node may mask the subsequent detection of a second fault on a non-redundant node.

Figure 9.14 illustrates the first possibility.

In this example, gate G3 is included to eliminate the static hazard

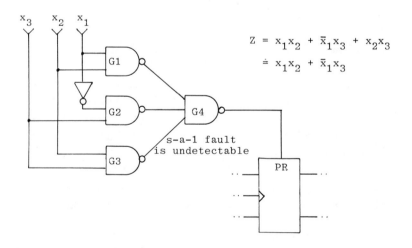

$$Z = x_1x_2 + \bar{x}_1x_3 + x_2x_3$$
$$\doteq x_1x_2 + \bar{x}_1x_3$$

FIGURE 9.14 Undetectable Fault Reintroducing A Hazard Condition

possibility when switching from the (x_1x_2) term to the $(\overline{x}_1x_3)$ term, i.e., G3 provides the ''bridging'' term. From a Boolean point of view, however, the output of G3 is logically redundant. Unfortunately it is not possible to propagate the effect of a s-a-1 fault on the output of G3 through gate G4. (The conditions to set G3 output to 0 are inconsistent with those required to set G1 and G2 outputs to 1. Hence a sensitive path through G4 cannot be established.) What this means is that the circuit will continue to operate correctly but with the possibility of a static-1 hazard condition (negative glitch) on G4 output as x_1 changes from 1 to 0 with x_2 and x_3 held at 1. The width of this pulse will be determined by the difference in the signal propagation paths and may or may not be wide enough to preset the flip-flop. It is important therefore that direct observation of the output of G3 is made possible so that the s-a-1 fault can be tested.

The second possibility is that a non-detectable fault on a redundant node can mask the detection of a normally-detectable fault. An example of this is shown in Figure 9.15.

The fault α s-a-1 is undetectable at z_1 because of the incompatible requirements for the values of x_1, x_2 and x_3.

The fault β s-a-0 is detected at z_1 by the test input $x_1 = 1$, $x_2 = 1$, $x_3 = 0$. (This is the only test for this fault.) The presence of α s-a-1, however, prevents detection of β s-a-1.

Guideline 3 Keep analog and digital circuits apart.

The tester requirements for analog circuits are quite different from those for digital circuits and it is useful to keep the two types of circuit separate, even if they exist on the same PCB. What this means is that analog signals that are inputs to analog-to-digital converters, should be brought out for observation prior to their conversion. Similarly, digital inputs to on-board digital-to-analog converters should be observable as digital signals. In this way, the analog and digital sections can be tested separately and with different test equipment if necessary.

Guideline 4 Partition a large circuit into smaller circuits

For an SSI/MSI-based PCB containing n devices, a rule of thumb for test-pattern generation is that the amount of effort to produce the tests is_

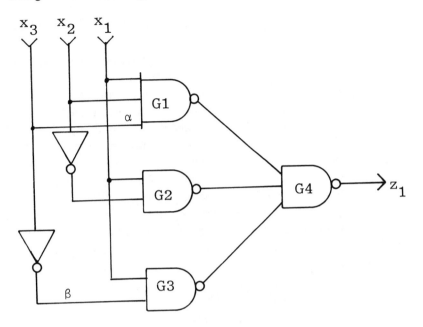

FIGURE 9.15 Fault Masking

proportional to n^3. If the circuit can be partitioned into two sub-circuits (for test purposes at least), then the amount of effort is reduced correspondingly. (For $n = 100$, the reduction in effort for two sub-circuits each containing 50 devices is $(50^3 + 50^3) : (100^3)$ i.e., roughly one quarter of the effort.

Partitioning of the circuit can be achieved by incorporating facilities to isolate and control clock lines, reset lines and possibly even power-supply lines. Alternatively, one section of a circuit can be separated from another by means of tristate buffers.

Guideline 5 Avoid asynchronous logic

Asynchronous logic is logic employing stored-state devices (in the form of latches) and global feedback but whose state-transition is governed solely by the sequence of changes on the primary inputs. There is no system clock to determine when the next state of the circuit will be established.

The advantage of asynchronous circuits is speed of operation. The speed at which state-transitions occur is limited only by the propagation delay of the gates and interconnects. In this respect, therefore, the design of asynchronous logic is more difficult than synchronous (clocked) logic and must be carried out with due regard to the possibility of races. Design techniques to eliminate race conditions do exist—the problem is that the presence of faults may reintroduce the race. Providing test patterns for such circuits can prove very difficult, particularly if the outcome of the race is not deterministic, i.e., in the critical race situation. Not only is the circuit difficult to test but the possibility of non-deterministic behavior also poses problems during the simulation stage. Overall therefore, synchronous logic is generally preferable to asynchronous logic.

Guideline 6 Make initialization easy

The importance of initialization was stressed in earlier chapters. Ideally it should be possible to set every stored-state device on the board into a known start state. What sometimes happens in practice is that individual preset or clear lines are tied to V_{cc} through a pullup resistor. Figure 9.16 illustrates some of the problems of, and solutions to, this particular requirement.

Other problems of initialization can occur if the state of one device is dependent on the state of another, as in an asynchronous ripple counter, for example. If independent control of the flip-flop in the counter is not provided, initialization can only be achieved by clocking the counter until a particular state is established and identified by the tester.

It is also important to initialize at least some of the contents of RAM devices.

Guideline 7 Provide facilities to break feedback paths

As with the previous guideline, the importance of this facility was stressed many times in earlier chapters. Feedback loops can be broken and controlled by a variety of techniques, as shown in Figure 9.17.

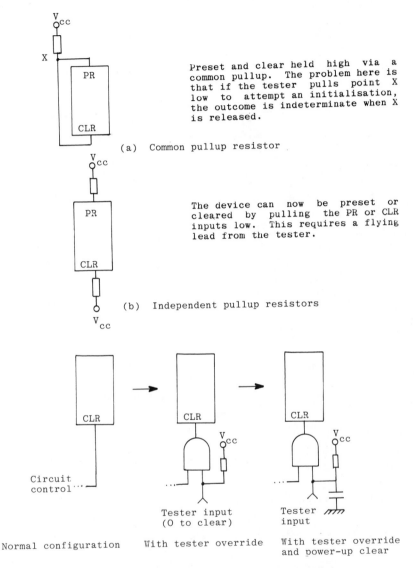

Preset and clear held high via a
common pullup. The problem here is
that if the tester pulls point X
low to attempt an initialisation,
the outcome is indeterminate when X
is released.

(a) Common pullup resistor

The device can now be preset or
cleared by pulling the PR or CLR
inputs low. This requires a flying
lead from the tester.

(b) Independent pullup resistors

Circuit
control

Tester input
(0 to clear)

Tester
input

Normal configuration   With tester override   With tester override
                                                     and power-up clear

(c) Tester override and power-up clear facilities

FIGURE 9.16 Initialization of Stored-State Devices

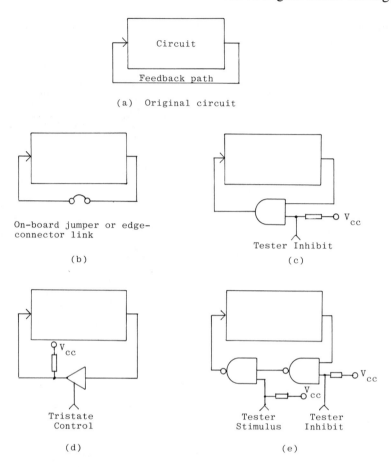

(a)   Original circuit

On-board jumper or edge-
connector link

(b)

Tester Inhibit

(c)

Tristate
Control

(d)

Tester   Tester
Stimulus   Inhibit

(e)

FIGURE 9.17 Techniques for Breaking and Controlling Feedback Paths

Guideline 8   Avoid monostables

There are various testing problems associated with the use of monostables in logic circuits.

The first is that direct observability of the monostable output is necessary if the period of the monostable is to be tested. If it is not possible to make the output directly observable then the test programmer will have to resort to a clip.

The second problem is that even if the output is observable, the period may be too fast for the tester. This problem can be solved either by the use of a follow-on latch to catch the pulse or by a capacitor clip over the monostable capacitor to lengthen the period. These techniques are illustrated in Figure 9.18.

Conversely, a monostable may have a long period (of the order of mS or even seconds). The problem now is not how to measure the width

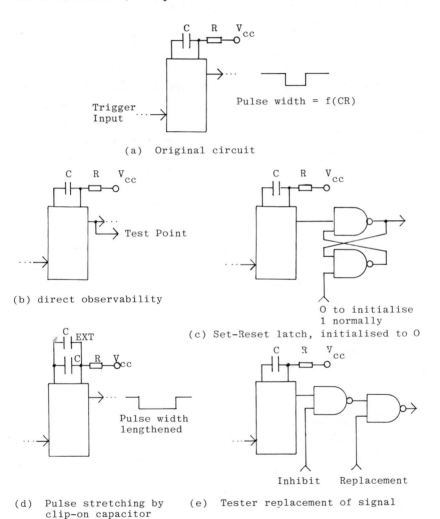

(a)   Original circuit

(b) direct observability

(c) Set-Reset latch, initialised to 0

(d)   Pulse stretching by   (e)   Tester replacement of signal
      clip-on capacitor

FIGURE 9.18 Techniques for Handling Monostables

of the pulse but how to shorten the pulse so that the period does not slow the testing speed. This can be achieved by a resistive clip over the monostable resistor.

Finally, if the use of a monostable cannot be avoided, then direct access to the reset line is important.

Guideline 9 Avoid ''singular'' components

Singular components are onboard components that are specially selected or adjusted to suit conditions on a particular board. They are commonly termed ''select-on-test'' and ''adjust-on-test'' components. Examples are: multi-turn locking potentiometers or tapped delay lines. The problem with such components is that it is often difficult to establish a standard test or, alternatively, that the test is more complex than need be. For example, a test program can include an automatic test to measure a fixed delay but will have to resort to a program loop and the use of oscilloscope to observe a delay whose value can vary from board to board.

Guideline 10 Take special care of ROMs

ROM devices are often used as replacements for combinational circuits implemented with basic gates. They are also used, with feedback, to realize a particular sequential behavior. From a testing point of view, however, there are other considerations.

The first is that the ROM may not contain the precise set of 0s and 1s necessary to enable the creation of a particular sensitive path some-where else in the circuit. If this is the case, then the test programmer may be forced to set the ROM output lines to their tristate level (by control-ling the ENABLE pin) and then clipover the top of the ROM to provide tester access to the ROM output pins.

This is shown in Figure 9.19.

In Figure 9.19(a), the ENABLE line has been tied directly to 0v, i.e., the ROM is permanently enabled. This means that the tester is unable to control the line. Figures 9.19(b) and (c) show two ways of allowing the tester access to the line. In Figure 9.19(b), the input is pulled low, whereas in Figure 9.19(c) it is pulled high. The latter scheme has more immunity to noise but requires more components.

Control of the ENABLE line and clipping of the device is important

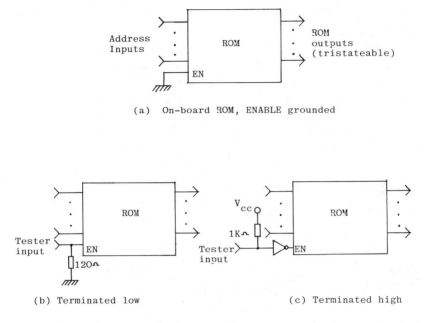

(a)   On-board ROM, ENABLE grounded

(b) Terminated low   (c) Terminated high

FIGURE 9.19 Terminated ROM Enable Line

from another point of view. The contents of a ROM device are sometimes changed as the board undergoes revision. It is sensible therefore to
reduce the dependency of the test program on the ROM contents in order
to prevent major modifications to the test program.

Guideline 11 Terminate all unused inputs and tristateable or open-
            collector outputs

From a design point of view, unused inputs to logic devices should
always be terminated to V_{cc} or 0v through a suitable resistor. This is
done to remove the risk of noise pickup on inputs left floating.

Termination of unused inputs is also important for testing purposes
in order to allow possible tester control of the behavior of a device (as
discussed in previous guidelines). It is important, however, that termination is via a resistor so that the tester can pull the terminated node high
or low.

Used and unused outputs from tristate or open-collector devices should also be terminated with a pull-up resistor to prevent inconsistent logic values leading to inconsistent signatures.

If the design of the system is such that a series of bus devices are assembled on different boards and connection from one device to another is via a back-plane bus, then it is common to find unterminated tristateable outputs. (The pull-up resistors are usually block mounted on a bus termination board.) For this situation, a special adaptor will be required to test the board if the tester does not contain pull-up or pull-down facilities.

One way to avoid this problem is to include switchable pull-up resistors for the bus outputs on each board.

9.3.2 Aids to Test-Application and Fault-Finding

The previous eleven guidelines related mostly to problems associated with the generation of suitable test patterns. This section continues with the guidelines that relate more to problems associated with applying the tests and to carrying out diagnostic procedures in the event of failure.

Guideline 12 Locate equivalent faults to the same IC package

If fault diagnosis is primarily by correlation of the circuit response with the entries in a fault dictionary, then it is useful to group equivalent faults onto the same integrated-circuit device. In this way, there is no need to attempt a further refinement of the diagnosis by means of a guided probe. Figure 9.20(a) contains an example of this—actually the example used in Chapter 5. (See Figure 5.1.)

In this circuit, it was discovered that a s-a-1 fault on c_1 was tested by the same set of tests as for the equivalent pair c_3 s-a-0 and c_6 s-a-1 around the inverter. If there is any choice therefore, the entire circuit should be implemented with gates from the same quad 2-input NAND gate package (Figure 9.20(b)) rather than with three NAND gates from one package and one inverter from another. (See Fig. 9.17(c).)

In practice, there is still the need to identify the fault source down to the actual node in order to eliminate the printed-circuit track as the cause of the fault. This requirement tends to invalidate the recommendation but, under certain circumstances, grouping equivalent faults can be

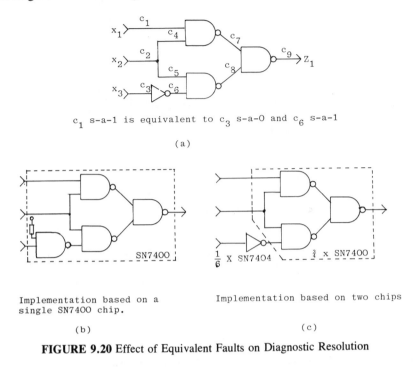

c_1 s-a-1 is equivalent to c_3 s-a-0 and c_6 s-a-1

(a)

Implementation based on a
single SN7400 chip.

Implementation based on two chips

(b)   (c)

FIGURE 9.20 Effect of Equivalent Faults on Diagnostic Resolution

useful, i.e., where it is known that a particular device is susceptible to failure. This is particularly true for gates whose outputs are wired together, either as a wired-OR or wired-AND.

Guideline 13 Allow tester control of clock lines

For circuits containing free-running on-board oscillators, it is useful to be able to replace the internal clock with one generated externally by the tester. In this way, the speed of operation of the circuit can be reduced, down to single-step operation if necessary. This has value not only for test application but also during the debug phase of test generation. Techniques for replacing internal clocks are shown in Figure 9.21.

Guideline 14 Avoid wired-OR, wired-AND junctions

Wired-OR, Wired-AND junctions present problems of ambiguity of fault diagnosis. If possible, they should be avoided. (See Figure 9.22.)

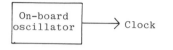

(a) Normal configuration

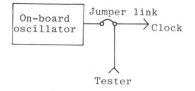

(b) Jumper or edge-connector link

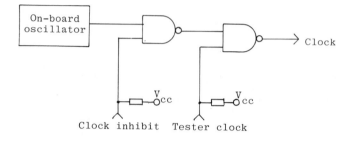

(c) Gate control

FIGURE 9.21 Techniques for Replacing Internal Clocks

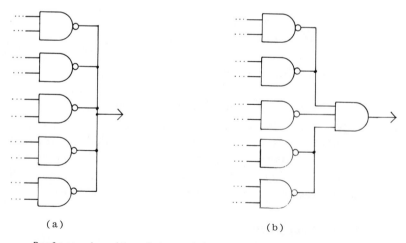

(a)   (b)

Replace circuits of type (a) with those of type (b)

FIGURE 9.22 Wired Junctions

It is also useful if gates whose outputs are wired together are confined to a small number of packages. In this way, if several gates are damaged because of a fault on the wired output connection, then only a few devices (preferably one) need be changed.

Guideline 15 Break up long counter chains

Figure 9.23(a) shows a 16-bit counter constructed from two 8-bit counters.

In order to test the counter fully, $2^{16} + 1 = 65537$ clock pulses should be applied. At a test rate of 50kHz, this represents a test time of approximately 1.3 seconds. Now consider the modified version shown in Figure 9.20(b). In this circuit, each 8-bit counter can be tested separately and the total test time is reduced to $2 \times (2^8 + 1) \times 20\mu s$, i.e.,

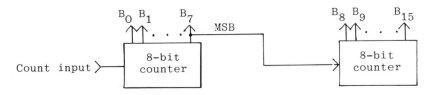

(a)   Normal configuration

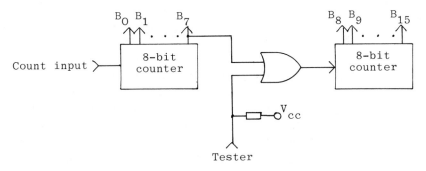

(b)   Independent count possibility

FIGURE 9.23 Breaking Up Long Counter Sequences

approximately 10 mS. The saving in time is important and is obviously valuable in its own right, but it is also useful if there are subsequent requirements to set the counter to a particular count for tests associated with other devices on the board. (In this respect, independent access to the pre-load facilities on each counter device is very useful.)

Guideline 16 Buffer edge-sensitive input signals

Many control signals to stored-state devices, such as clock signals, specify minimum rise or fall times. If these signals come direct from an edge-connector position, then it may be necessary to incorporate a buffer device (line driver or two inverters) in the interface between the board and tester to speed-up the rise and fall times of tester-generated signals. The need to do this is dependent on the slew rate of the driver amplifier in the tester. Effectively, what this guideline says is that high-speed devices should be kept away from the primary inputs and primary outputs.

Guideline 17 Leave room around devices

There may be a requirement to gain access to a device by means of a single-pin or multi-pin clip. If this is likely, then care should be taken to ensure that there is enough room to accommodate such a clip. Logic devices are not usually located too closely together, but occasionally discrete components such as pull-up resistors or decoupling capacitors are located very close to a device.

Guideline 18 Take care with board layout and construction

To help reduce operator fatigue when fault-finding with a guided probe over a range of different board types, it is helpful if device orientation is standard, i.e., if pin 1 is always top-left-hand corner, say. Also, mis-probes can occur as the operator moves from a 14-pin device to a 16-pin device, although it is difficult to suggest a solution to this problem. Possibilities are to mark all 16-pin devices with a color marker or to concentrate such devices into one or two columns on the board.

There should also be clear identification of the board number and modification level, of the grid reference on the board, and of any discrete components that are identified by a special identifier (R1, R2, etc.) rather than by their grid reference.

For boards that are to be diagnosed by guided probe, it is helpful if all devices and printed-circuit tracking are visible from a single side of the board.* The visibility of devices is necessary for conventional probing and the visibility of track is necessary for current-sensing probes. (In this context, it is not possible to follow copper track that is routed beneath an integrated circuit device. Similarly, plated-through holes can create tracing problems.) For any form of contact probing, non-conducting protective coatings are to be avoided.

Finally, on-board test points that are not brought out to edge-connector positions should be gathered close together to simplify the interface requirement of flying leads. This can be done by using a dummy integrated-circuit socket or by centralizing all test points onto a series of 0.1″ pitch stake pins or wire-wrap pins. Access is then via a single harness terminated with a 14-pin plug or socket.

Guideline 19 Socket complex devices

There are two reasons for mounting complex devices in sockets. The first is that they can be easily removed during the application of the tests, if required, and the second is that they can be easily replaced if found to be faulty. Unsoldering 40-pin or 64-pin devices is not an attractive task! The disadvantage with the use of sockets is the increased risk of failure caused by a bad contact between the device pin and the socket.

Guideline 20 Provide a short-circuit link to check board alignment on the tester

It is good practice for the operator to check the alignment and orientation of the board before running the program, but it is helpful if the program itself can carry out a check before applying power. One way to do this is to make use of a pair of unused edge-connector positions and to short-circuit between these positions on the board. The test program can then test that the short-circuit exists before continuing with the program. (See Figure 9.24.)

_____

*Although this is inconsistent with the modern practice of multi-layer, densely-populated boards.

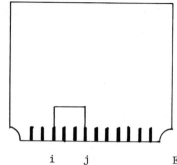

i   j   Edge-connector fingers

Edge-connector fingers, i and j,
have been shorted together on the
board to allow the test programmer
to include a test for correct
alignment of the board prior to the
application of power. In this way,
the possibility of damage (to board
or tester) due to mis-alignment is
considerably reduced.

FIGURE 9.24 Automatic Testing of Board Alignment

Guideline 21 Be wary of the interface requirements

The standard interface between a board and the tester consists of the
primary input, primary output stimulus/response interface, plus power
supply distribution. In addition, however, we have noted that there may
be many other specialized interface requirements such as "flying lead"
probes (single or multi-pin), buffer devices, clipped components, etc.
Although each of these facilities is provided to ease a particular testing
problem, they represent an additional complication from the user's point
of view and, in a distributed testing environment, from a cost and
logistics point of view. The leads may also represent a considerable
capacitative load on the board. The general engineering maxim of
"keep it simple" applies just as much here as elsewhere.

Guideline 22 Avoid mixed logic on the same board

Different logic families require different reference values on driver/
sensor pins and, usually, different power-supply levels. The use of two

or more logic families on the same board can therefore cause various complications to the interface. One solution, if mixed logic is unavoidable, is to ensure that all primary inputs and primary outputs are at least TTL compatible.

Guideline 23 Limit device fanout

If a device output is being used to drive its maximum load, then the addition of a guided probe on the node may just be enough to cause degradation of the signal. To avoid this problem, high-fanout devices should be limited to drive one less than their design maximum.

Guideline 24 Provide clear design engineering documentation

Last, but not least, the design engineer should provide clear documentation to support the design of the board. In particular, the documentation should include the following items.

(i)   A complete functional specification for the board, including an explanation of signal names where these are meaningful.
(ii) Timing diagrams and tolerances showing the results of control actions.
(iii) Details of any design workarounds and notification of any parts of the design that may be modified in the future.
(iv) Clear logic diagrams together with identification of major and minor feedback paths.
(v)   A recommended test strategy.

9.4 Concluding Remarks on Testable Design

In this chapter, we have looked at methods for improving the testability of logic circuits. Some of the methods are formal whereas others are no more than commonsense rules. In practice however, all the methods have the same objective: to reduce the costs of testing. An empirical relationship that has been used for estimating the cost of finding a faulty device is that the cost will increase by a factor of 10 as fault-finding moves from one level to the next, i.e., the cost of finding a faulty device at:

1. device level (goods inward)   1 unit of cost
2. board level (board test)   10 units of cost
3. system level (system test)   100 units of cost
4. field level (field service)   1000 units of cost

Over the past few years, however, the cost of testing and fault-finding at systems and field level has risen even higher than the factor of 10 above and the situation now looks more like that shown in Figure 9.25.

In effect, the costs are neither predictable nor controllable and the only solution is to place more emphasis on testability. This means acceptance of testability as a design constraint and an understanding of what the rules are and why they are necessary. This chapter has presented the rules; the previous eight chapters have attempted to show why they are necessary.

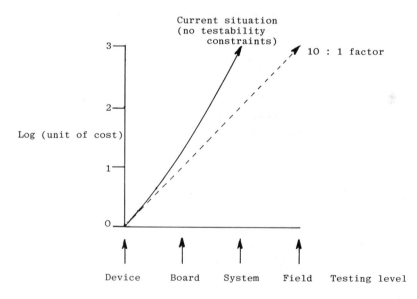

FIGURE 9.25 Costs of Testing at Various Levels

Chapter 10
A Guide to Further Reading

As stated in the preface and in the title of the book, this book is designed to be an introduction to the subject of testing digital boards. It is important, therefore, to know where specific subjects can be pursued further and which journals or conferences should be scanned. This chapter presents a guide to sources of further information. Note, however, that the lists of journals, conferences, textbooks, and references on specific topics given here are not exhaustive. There are two reasons for this. The first is that an exhaustive list—of references in particular—would be so long as to lose much of its value; the second is that the diligent researcher can always work back through reference lists attached to the references given here. There is no point, therefore, in giving an exhaustive list in this book.

The other general comment about the sources of information mentioned in this chapter is that, to a certain extent, the list is subjective. This means that the entries are the sources that the author has found to be useful over some ten years' involvement in the subject.

The overall structure of the chapter is to present general sources of information first, followed by lists of specific references for defined topics. The general sources are broken down into textbooks, journals, conferences, government and university agencies, information retrieval services, and bibliographies. The specific references are organized by topic headings and identified by the chapter in the book in which the topic is described.

In general, lists are organized in chronological order starting with the earliest reference. Also, each reference is given a unique number. The chapter concludes with a separate main author/journal title/conference title index. This indexing information is not duplicated in the main index to the book.

10.1 Sources of Information

10.1.1 Textbooks

1. H.Y. Chang, E.G. Manning and G. Metze. *Fault Diagnosis of Digital Systems*. Wiley-Interscience, 1970. (Reprinted 1974 by Krieger Publishing Co., New York).

2. Z. Kohavi. *Switching and Finite Automata Theory*. McGraw-Hill, 1970, (Chaps. 8 and 13).

3. A.D. Friedman and P.R. Menon. *Fault Detection in Digital Circuits*. Prentice-Hall, 1970.

4. M.A. Breuer (General Editor). *Design Automation of Digital Systems:* Vol. 1, *Theory and Techniques*. Prentice-Hall, 1972.

5. R. Knowles. *Automatic Testing: Systems and Applications*. McGraw-Hill, 1976.

6. S.C. Lee. *Digital Circuits and Logic Design*. Prentice-Hall, 1976, (Chaps. 7 and 8).

7. M.A. Breuer and A.D. Friedman. *Diagnosis and Reliable Design of Digital Systems*. Computer Science Press, 1976.

8. D.W. Lewin. *Computer-aided Design of Digital Systems*. Arnold, 1977, (Chap. 5).

9. S.C. Lee. *Modern Switching Theory and Digital Design*. Prentice-Hall, 1978, (Chaps. 6 and 10).

10. J.P. Roth. *Computer Logic, Testing and Verification*. Computer Science Press, 1980.

10.1.2 Journals

In the following list of journals, the first five (nos. 11–15) constitute the "core" journals for archival papers on testing. Most of the other journals mentioned carry occasional articles on testing, many of which are tutorial in nature. The last two journals mentioned (nos. 26 and 27) are the main European and American trade journals.

11. IEEE TRANSACTIONS ON COMPUTERS, formerly IRE/IEEE Transactions on Electronics Computers (changeover in 1967/68). Published monthly by IEEE, USA. Special issues on fault-tolerant computing (which embraces testing) have appeared as follows:

> November, 1971 (Vol. C-20)
> March, 1973 (Vol. C-22)
> July, 1974 (Vol. C-23)
> May, 1975 (Vol. C-24)
> June, 1976 (Vol. C-25)
> June, 1978 (Vol. C-27)
> June, 1980 (Vol. C-29)

12. COMPUTER. Published monthly by the IEEE Computer Society, USA.

13. JOURNAL OF DIGITAL SYSTEMS, formerly Journal of Design Automation and Fault-Tolerant Computing (changeover with Vol. 4, No. 1, spring, 1980). Published quarterly by Computer Science Press, USA.

14. DIGITAL PROCESSES. Published quarterly by Georgi Publishing Company, Switzerland.

15. IEE PROCEEDINGS PART E: COMPUTERS AND DIGITAL TECHNIQUES. Published quarterly by IEE, UK.

16. IEEE SPECTRUM. Published monthly by IEEE, USA.

17. IEEE PROCEEDINGS. Published monthly by IEEE, USA.

18. COMPUTING SURVEYS AND COMPUTING REVIEWS. Published quarterly by the Association for Computing Machinery (ACM), USA.

19. ELECTRONICS INTERNATIONAL. Published every two weeks by McGraw-Hill, USA.

20. ELECTRONICS DESIGN. Published every two weeks by Hayden Publishing Co., USA.

21. COMPUTER DESIGN. Published monthly by Computer Design Publishing Co., USA.

22. BELL SYSTEM TECHNICAL JOURNAL. Published ten times a year by the American Telephone and Telegraph Co., USA.

23. IBM JOURNAL OF RESEARCH AND DEVELOPMENT. Published every two months by IBM, USA.

24. IEE ELECTRONICS LETTERS. Published every two weeks by IEE, UK.

25. IERE RADIO AND ELECTRONIC ENGINEER. Published monthly by IERE, UK.

26. TEST, formerly Journal of ATE (changeover in 1979). Published every two months by Network, UK.

27. ELECTRONICS TEST. Published every month by Benwell Publishing Co., USA.

10.1.3 Conferences

The subject of testing usually occupies a prominent position in any conference concerned with the theory, design, or application of digital systems. Several conferences exist, however, for which testing is a main theme, if not the only theme. These conferences are listed below.

28. IEEE Computer Society TEST CONFERENCE, previously known as the Semiconductor Test Symposium (changeover in 1979). Held annually in October in the USA at Cherry Hill, New Jersey.

29. IEEE AUTOTEST Conference, held annually in November in the USA.

30. IEEE Computer Society SYMPOSIUM ON FAULT-TOLERANT COMPUTING. Held annually, usually but not always in the USA, in June. (Note: the special issues of the IEEE Transactions on Computers are based on the papers presented at these symposia.)

31. Joint IEEE/ACM DESIGN AUTOMATION Conference, held annually in June in the USA.

32. AUTOMATIC TESTING Conference and Exhibition, organized annually by Network, UK, and held in Europe.

33. IEE/IERE AUTOMATIC TEST SYSTEMS. An occasional series of conferences held biannually in the UK.

34. MICROTEST, organized by the UK Society of Electronic and Radio Technicians and inaugurated in April, 1979.

10.1.4 Government and University Agencies

There are several central agencies for distributing university or industry-derived reports. These provide a valuable service in making such reports generally available. The principal agencies are:

35. NATIONAL TECHNICAL INFORMATION SERVICE (NTIS). The NTIS is the main distribution agency for reports emanating from industrial, university, and military establishments in America. NTIS reports are usually allocated an "AD" number, and the reports are available in either photocopy or microfiche form. Full details can be obtained from:

> NTIS,
> 5285, Port Royal Road,
> Springfield,
> Virginia 22151.
> USA

Alternatively, AD reports (microfiche form only) can be obtained through the following European agencies:

> Space Documentation Service,
> European Space Agency,
> 114 Avenue Charles de Gaulle,
> 92200 Neuilly-sur-Seine,
> France

> Technology Reports Centre (DOI)
> Station Square House,
> St. Mary Cray,
> Orpington,
> Kent BR5 3RF
> England

36. SCIENTIFIC AND TECHNICAL AEROSPACE REPORTS (STAR). These reports are published by the American National

Aeronautics and Space Administration (NASA) and full details are available from:

NASA Scientific and Technical Information Facility,
P.O. Box 8757,
Baltimore/Washington International Airport,
Maryland 21240
USA

37. UNIVERSITY MICROFILMS. Photocopies of American University theses. Full details from:

University Microfilms,	or	University Microfilms,
30 Nzeeb Road,		Tylers Green,
Ann Arbor,		High Wycombe,
Michigan, USA		England

10.1.5 Information retrieval services

A major information retrieval service for electronic engineers is the joint IEE/IEEE Information Service in Physics, Electrotechnology and Control (INSPEC). As the title suggests, this service covers a wide range of electronic subjects. The two publications covering digital testing are:

38. COMPUTER AND CONTROL ABSTRACTS, Science Abstract Series C, published monthly.

39. CURRENT PAPERS IN COMPUTERS AND CONTROL, published monthly.

The second of these publications contains details of the author, title, and place of publication only. The Computer and Control Abstracts, however, also includes the author's abstract.

Further details about the INSPEC service can be obtained from:

Fulfillment Manager,	INSPEC Marketing Department,
IEEE,	IEE,
445 Hoes Lane,	Savoy Place,
Piscataway,	London WC2R OBL
New Jersey 08854	England
USA	

(The reader who is not too familiar with the INSPEC service may like to read the short but informative article by A.A. McKenzie, entitled "Information retrieval: one wheel is enough," published in IEEE Spectrum, Vol. 10, No. 4, April 1973, pp. 55–56.)

Another major information retrieval system is:

40. SCIENCE CITATION INDEX

The Science Citation Index is based on approximately 3000 technical publications and is published in three sections as follows:

(i) Citation Index. This is an index of authors, both prime and secondary, who have been cited in the reference sections of the technical publications.

(ii) Source Index. This is an index of all listed authors of the technical publications.

(iii) Permuterm Subject Index. This is an index of keywords taken from the titles of the technical publications.

The Citation Index and Source Index are arranged alphabetically by author surname, whereas the Subject Index is arranged alphabetically by keyword.

The three sections are collectively known as the Science Citation Index, which is published in the following way:

(i) every two months;

(ii) every year, as a single cumulative issue to replace the 2-monthly issues;

(iii) every five years, as a single cumulative issue to replace the yearly issues.

Further details can be obtained from:

iSi	or	iSi (European Branch)
325 Chestnut Street,		132 High Street,
Philadelphia,		Uxbridge,
PA 19106		Middlesex UB8 1DP
USA		UK

It is also worth noting that iSi provides a "Current Contents" service. Effectively, this is a service to provide copies of the contents pages of a subset of the technical publications scanned by iSi. The publications are broken down into seven main fields; the field containing electronic engineering is "Engineering and Technology."

(A useful description of the Science Citation Index is contained in the article written by E. Garfield, entitled "Science Citation Index—a new dimension in indexing" published in Science, Vol. 144, No. 3619, May 8, 1964, pp. 649–654.)

10.1.6 Bibliography

Review papers containing extensive bibliographies appear quite regularly in the literature and some are mentioned in the specific lists later in this chapter. An extensive bibliography of technical publications in the general area of computer-aided design of digital systems has been compiled and is regularly updated. The details are as follows:

41. V.M. VanCleemput. *Computer-aided design of digital systems—a bibliography.* Vol. 1, 1960–1974; Vol. 2, 1975–1976; Vol. 3, 1976–1977; Vol. 4, 1977–1979.

The bibliography is published by:

Computer Science Press Inc.,
9125 Fall River Lane,
Potomac,
Maryland 20854
USA

10.2 Specific references and bibliography

This section identifies references on specific topics introduced throughout the earlier chapters of the book. The following abbreviations are used for the major journals and conferences:

DA	IEEE Design Automation Conference
DAFTC	Journal of Design Automation and Fault-Tolerant Computing (now Digital Systems)
FTCS	IEEE Fault Tolerant Computing Symposium
IEEETC	IEEE Transactions on Computers

IEEETEC   IEEE Transactions on Electronic Computers
STC   IEEE Semiconductor Test Conference
TC   IEEE Test Conference

10.2.1 Chapter 1

Fault-effect modelling (stuck-at, bridging, multiple, intermittent)

42. J.W. Gault, J.P. Robinson, and S.M. Reddy. *Multiple Fault Detection in Combinational Networks*. IEEETC, Vol. C-21, 1972, pp. 31–36.

43. M.A. Breuer. *Testing for Intermittent Faults in Digital Circuits*. IEEETC, Vol. C-22, 1973, pp. 241–246.

44. S. Kamal and C.V. Page. *Intermittent Faults: A Model and a Detection Procedure*. IEEETC, Vol. C-23, 1974, pp. 713–719.

45. K.C.Y. Mei. *Bridging and Stuck-at Faults*. IEEETC, Vol. C-23, 1974, pp. 720–726.

46. A.D. Friedman. *Diagnosis of Short Faults in Combinational Circuits*. Proc. 3rd FTCS, 1975, pp. 95–99.

47. S. Kamal. *An Approach to the Diagnosis of Intermittent Faults*. IEEETC, Vol. C-24, 1975, pp. 461–467.

48. S.S. Yau and S.C. Yang. *Multiple Fault Detection for Combinational Logic Circuits*. IEEETC, Vol. C-24, 1975, pp. 233–242.

49. J. Hlavicka and E. Kottek. *Fault Models for TTL Circuits*. Digital Processes, Vol. 2, 1976, pp. 169–180.

50. J.T. Easterbrook and R.G. Bennetts. *Failure Mechanisms in Logic Circuits and Their Related Fault Effects*. Proc. IEE Conference on New Developments in Automatic Testing, Nov. 1977, pp. 44–47.

51. C.H. Lin and S.Y. Su. *Feedback Bridging Faults in General Combinational Networks*. Proc. 8th FTCS, 1978, p. 211.

52. J. Galiay, Y. Crouzet, and M. Vergniault. *Physical Versus Logical Fault Models in MOS LSI Circuits: Impact on Their Testability*. IEEETC, Vol. C-29, 1980, pp. 527–531.

53. K.L. Kodandapani and D.K. Pradhan. *Undetectability of Bridging Faults and Validity of Stuck-at-Fault Test Sets*. IEEETC, Vol. C-29, 1980, pp. 55–59.

54. M. Karpovsky and S.Y. Su. *Detection and Location of Input and Feedback Bridging Faults among Input and Output Lines.* IEEETC, Vol. C-29, 1980, pp. 523–527.

10.2.2 Chapter 2

(i) Fault collapsing

55. E.J. McCluskey and F.W. Clegg. *Fault Equivalence in Combinational Logic Networks.* IEEETC, Vol. C-20, 1971, pp. 1286–1293.
56. D.R. Schertz and G. Metze. *A New Representation for Faults in Combinational Digital Circuits.* IEEETC, Vol. C-21, 1972, pp. 858–866.

(ii) Fault-effect reconvergence

57. D.B. Armstrong. *On Finding a Nearly Minimal Set of Fault Detection Tests for Combinational Logic Nets.* IEEETEC, Vol. EC-15, 1966, pp. 66–73.
58. R.G. Bennetts and E.H. Manolev. *A Procedure for Determining the Fault-cover for Tests for Combinational Circuits.* Proc. 3rd Seminar on Applied Aspects of Automata Theory, Varna, Bulgaria, June 1975, pp. 412–421.

10.2.3 Chapter 3

(i) D-algorithm

59. J.P. Roth. *Diagnosis of Automata Failures: A Calculus and a Method.* IBM Journal Research and Development, Vol. 10, No. 7, July 1966, pp. 278–291.
60. J.P. Roth, W.G. Bouricius, and P.R. Schneider. *Programmed Algorithms to Compute Tests to Detect and Distinguish between Failures in Logic Circuits.* IEEETEC, Vol. EC-16, 1967, pp. 567–580.
61. G.R. Putzolu and J.P. Roth. *A Heuristic Algorithm for the Testing*

of Asynchronous Circuits. IEEETC, Vol. C-20, 1971, pp. 631–647.

62. H. Kubo. *A Procedure for Detecting Test Sequences to Detect Sequential Circuit Failures.* NEC Journal Research and Development, No. 12, 1968, pp. 66–78.

(ii) LASAR algorithm

63. J.J. Thomas. *Automated Diagnostic Test Program for Digital Networks.* Computer Design, Vol. 10, No. 8, Aug. 1971, pp. 63–67.

64. K.R. Bowden. *A Technique for Automatic Test Generation for Digital Circuits.* Proc. IEEE Intercon, 1975, Session 15, pp. 1–5.

10.2.4 Chapter 4

(i) Development of simulators: general references

65. *The LAMP System.* Special issue of Bell System Technical Journal, Vol. 53, No. 8, 1974.

66. GenRad. *Simulator-based Interactive Software Aids to Test Generation.* Publication JNTSD3-875.

67. *Digital System Simulation.* Special issue of IEEE Computer, March 1975.

68. S.G. Chappell, P.R. Menon, J.F. Pellegrin, and A.M. Schowe. *Functional Simulation in the LAMP System.* Journal DAFTC, Vol. 1, 1977, pp. 203–215.

69. Hewlett-Packard Co. *Modelling and Simulation for Digital Testing.* Application Note 210-1, Jan. 1977.

70. D. Blunden, A.H. Boyce, and G. Taylor. *Logic Simulation.* The Marconi Review, Vol. 50, 1977, No. 206, pp. 157–171 (Part 1); No. 207, pp. 236–253 (Part 2).

71. B. Childs and B.M. Kasindorf. *Advanced Modelling Technique Provides Capability to Simulate/Test Pcbs Containing LSI.* Proc. IEEE Nepcon, Feb. 1978, pp. 388–398.

72. W.A. Johnson, R.W. Rozeboom, and J.J. Crowley. *Digital Circuit Test Analysis and Failure Diagnosis using the TITAN System.* Journal DAFTC, Vol. 1, 1977, pp. 287–309.

73. Y.H. Levendel and P.R. Menon. *Unknown Signal Values in Fault Simulation.* Proc. 9th FTCS, 1979, pp. 125–128.

74. E. Kjelkemid and O. Thessen. *Methods of Modelling Digital Devices for Logic Simulation.* Proc. 16th DA, 1979, pp. 235–241.

75. P. Wilcox. *Digital Logic Simulation at the Gate and Functional Level.* Proc. 16th DA, 1979, pp. 242–248.

76. M.A. Breuer and A.D. Friedman. *Functional Level Primitives in Test Generation.* IEEETC, Vol. C-29, 1980, pp. 223–235.

(ii) Parallel simulation

77. S. Seshu and D.N. Freeman. *The Diagnosis of Asynchronous Sequential Switching Systems.* IRE Trans. Electronic Computers, Vol. EC-11, 1962, pp. 459–465.

78. S. Seshu. *On an Improved Diagnosis Program.* IEEETEC, Vol. EC-14, 1965, pp. 69–76.

79. P.L. Flake, G. Musgrave, and I. White. *A Digital System Simulator: HILO.* Digital Processes, Vol. 1, No. 1, 1975, pp. 39–53.

80. S.A. Szygenda and E.W. Thompson. *Modelling and Digital Simulation for Design Verification and Diagnosis.* IEEETC, Vol. C-25, 1976, pp. 1242–1252.

(iii) Deductive simulation

81. D.B. Armstrong. *A Deductive Method for Simulating Faults in Logic Circuits.* IEEETC, Vol. C-21, 1972, pp. 464–471.

82. H.Y. Chang and S.G. Chappell. *Deductive Techniques for Simulating Logic Networks.* IEEE Computer, Vol. 8, No. 3, March 1975, pp. 52–59.

83. F. Wang, E. Lowe and F. Angeli. *Design of a Functional Test Generator with a Functional Deductive Simulator for Digital Systems.* Proc. IEEE Autotest Conference, Nov. 1978, pp. 134–142.

(iv) Concurrent simulation

84. E.G. Ulrich and T.E. Baker. *Concurrent Simulation of Nearly Identical Digital Networks.* IEEE Computer, April 1974, pp. 39–44.

85. D.M. Schuler, T.E. Baker, S.P. Bryant, and E.G. Ulrich. *A Computer Program for Logic Simulation, Fault Simulation and the Generation of Tests for Digital Circuits*. Proc. 8th AICA Congress on Simulation of Systems, Aug. 1976, pp. 453–459.

86. M. Abramovici, M.A. Breuer, and K. Kumar. *Concurrent Fault Simulation and Functional Level Modelling*. Proc. 14th DA, 1977, pp. 128–137.

87. D.M. Schuler et al. *A Program for the Simulation and Concurrent Fault Simulation of Digital Circuits Described with Gate and Functional Models*. Proc. STC, 1979, pp. 203–207.

10.2.5 Chapter 5

Transition count signatures

(For papers on CRC signatures, see the papers listed under Signature Analysis in Section 10.2.8.)

88. J.P. Hayes. *Transition Count Testing of Combinational Logic Circuits*. IEEETC, Vol. c-25, pp. 613–620.

89. J.K. Skilling. *Signatures Take a Circuit's Pulse by Transition Counting or PRBS—but Watch Those Loops*. Electronic Design, Vol. 28, No. 3, Feb. 1, 1980, pp. 65–68.

10.2.6 Chapter 6

(i) ATE/ATS: general references

90. K. To and R.E. Tulloss. *Automatic Test Systems*. IEEE Spectrum, Vol. 11, No. 9, Sept. 1974, pp. 44–52.

91. E.A. Torrero. *ATE: not so easy*. IEEE Spectrum, Vol. 14, No. 4, April 1977, pp. 29–34.

92. R.E. Allan. *New Philosophies for Portable Digital Instruments*. IEEE Spectrum, June 1978, pp. 32–37.

93. D. Mennie. *Brainer On-site Trouble-Shooting Tools Dig Out Component-Level Faults*. Electronic Design, Vol. 27, No. 21, Oct. 11, 1979, pp. 74–79.

(ii) Device testing

94. F. van Veen. *An Introduction to IC Testing*. IEEE Spectrum, Vol. 8, No. 12, Dec. 1971, pp. 28–37.

(iii) In-circuit testing

95. D.W. Raymond. *Component by Component Testing of Digital Circuit Boards*. Computer Design, Vol. 19, No. 4, April 1980, pp. 129–137.

(iv) Testing economics

96. Fluke. *The Economics of Logic Board Testing*. Brochure.
97. B. Davis. *Optimising the ATE Test Mix*. Electronics Production, Jan. 1980.
98. D. Moralee. *Economics of Using ATE*. IEE Electronics and Power, Vol. 26, No. 2, Feb. 1980, pp. 176–182.

10.2.7 Chapter 8

(i) Testing microprocessor devices/boards: general references

99. R. Ungerman. *Microcomputer Testing Strategies*. Proc. IEEE Symposium on Circuits and Systems, April 1975, pp. 363–365.
100. A.C.L. Chiang. *Testing Schemes for Micro-processor Chips*. Computer Design, Vol. 14, No. 4, April 1975, pp. 87–92.
101. D. Hackmeister and A.C.L. Chiang. *Microprocessor Test Techniques Reveal Instruction Pattern Sensitivity*. Computer Design, Vol. 14, No. 12, Dec. 1975, pp. 81–85.
102. E.R. Hnatek. *Checking Microprocessors*. Electronic Design, Vol. 23, No. 22, Oct. 25, 1975, pp. 102–105.
103. W. Luciw. *Can a User Test LSI MPs Effectively?* IEEE Trans Manufacturing Technology, Vol. MFT-5, No. 1, March 1976, pp. 21–23.
104. A.C.L. Chiang and R. McCaskill. *Two New Approaches Simplify Testing of Microprocessors*. Electronics International, Vol. 49, No. 2, Jan. 22, 1976, pp. 100–105.
105. J. Barnes and B. Bergquist. *Unite MP Hardware and Software*. Electronic Design, Vol. 24, No. 7, March 29, 1976, pp. 74–76.

106. B. Schusheim. *A Flexible Approach to Microprocessor Testing.* Computer Design, Vol. 15, No. 3, March 1976, pp. 67–72.

107. R.E. Anderson. *Test Methods Change to Meet Complex Demands.* Electronics International, Vol. 49, No. 8, April 15, 1976, pp. 125–128.

108. E.G. Foley and A.H. Firman. *Testing Microcomputer Boards Automatically.* Computer Design, Vol. 15, No. 12, Dec. 1976, pp. 92–94.

109. A. Santoni. *Testers are Getting Better at Finding Microprocessor Flaws.* Electronics International, Vol. 49, No. 24, Dec. 23, 1976, pp. 57–66.

110. A.C.L. Chiang. *Test Schemes for Microprocessor Chips.* Computer Design, Vol. 15, No. 4, April 1976, pp. 125–128.

111. R. McCaskill. *Wring out 4-bit MP Slices with Algorithmic Pattern Generation.* Electronic Design, Vol. 25, No. 9, May 10, 1977, pp. 74–77.

112. D.H. Smith. *Exercising the Functional Structure Gives Microprocessors a Real Workout.* Electronics International, Vol. 50, No. 4, Feb. 17, 1977, pp. 109–112.

113. R.E. Anderson. *Board Testing in the 80's.* Proc. STC, 1979, pp. 7–15.

114. K. Ripley. *Testing Microprocessor-based Circuits.* Electronic Engineering, Vol. 49, No. 597, Oct. 1977, pp. 65–68.

115. W.G. Fee (General Editor). *Tutorial on LSI Testing.* IEEE Computer Society, 1977 (2nd edition). IEEE Catalog No. EHO 122-2.

116. S.R. Purks. *Flexibility for Testing Boards Containing LSI Components.* Proc. IEEE Electro 77, April 1977, Session 32, Paper No. 2.

117. J. Grason. *Testing Circuit Packs Containing LSI Components.* Proc. IEEE Electro 77, 1977, Session 32, Paper No. 1.

118. S.E. Scrupski. *Why and How Users Test Microprocessors.* Electronics International, Vol. 51, No. 5, March 2, 1978, pp. 97–104.

119. R.P. Capece. *Tackling the Very Large-Scale Problems of VLSI: A Special Report.* Electronics International, Vol. 51, No. 23, 1978, pp. 111–125.

120. P. Hansen. *An Advanced Test Strategy Development for Micro-computer Boards.* Proc. TC, 1979, pp. 354–359.

121. S. Runyon. *Testing LSI-based Boards: Many Issues, Many Answers.* Electronics Design, Vol. 27, No. 6, March 15, 1979, pp. 58–66.

122. GenRad. *Testing a Microprocessor Board on the GenRad 2225.* Application Note, 1979.

123. M.J. Chalkley. *Trends in VLSI Testing.* Proc. TC, 1979, pp. 3–6.

124. R.P. Davidson and N.R. Miller. *Microprocessor System Testing and Diagnostics.* Proc. 19th IEEE Compcon, 1979, pp. 105–110.

125. S.M. Thatte and J.A. Abraham. *Test Generation for Microprocessors.* IEEETC, Vol. C-29, 1980, pp. 429–441.

(ii) Device reliability and failure modes

126. E.R. Hnatek. *Microprocessor Device Reliability.* Microprocessors and Microsystems, Vol. 1, No. 5, June 1977, pp. 299–303.

127. B. Halil. *Microprocessor Failure Rate Predictions.* Microelectronics and Reliability, Vol. 17, No. 1, 1978, pp. 211–222.

(iii) Testing RAMs

128. J. Reese-Brown. *Pattern Sensitivity in Semiconductor Memories.* STC, 1972, pp. 33–46.

129. E.R. Hnatek. *4-Kilobit Memories Present a Challenge to Testing.* Computer Design, Vol. 14, No. 5, May 1975, pp. 117–125.

130. J.P. Hayes. *Detection of Pattern-sensitive Faults in Random Access Memories.* IEEETC, Vol. C-24, 1975, pp. 150–157.

131. Macrodata Corporation. *Selecting Test Patterns for 4K RAMs.* Application Note 139, April 1977.

132. S.M. Thatte and J. Abraham. *Testing of Semiconductor Random Access Memories.* Proc. 7th FTCS, 1977, pp. 81–87.

133. D.S. Suk and S.M. Reddy. *Test Procedures for a Class of Pattern-Sensitive Faults in Semiconductor RAMs.* IEEETC, Vol. C-29, 1980, pp. 419–429.

(iv) Alpha-particle radiation

134. T.C. May and M.H. Woods. *A New Physical Mechanism for Soft*

Errors in Dynamic Memories. Proc. 16th IEEE Reliability Physics Symposium, April 1978, pp. 33–40.

135. M. Brodsky. *Hardening RAMs Against Soft Errors*. Electronics International, Vol. 53, No. 8, April 24, 1980, pp. 117–122.

(v) Propagation delay faults

136. E.P. Hsieh, R.A. Rasmussen, L.J. Vidunas, and W.T. Davis. *Delay Test Generation*. Proc. 14th DA, 1977, pp. 486–491.

137. J. Shedletsky. *Delay Testing LSI Logic*. Proc. 8th FTCS, 1978, pp. 159–164.

138. J.D. Lesser and J. Shedletsky. *An Experimental Delay Test Generator for LSI Logic*. IEEETC, Vol. C-29, pp. 235–248.

(vi) Current-sensing probes

139. Hewlett-Packard. *Current Tracer 547A* and *Logic Pulser 546A*. Operating manuals.

140. P.C. Grace. *The Electronic Knife-Automatic Fault Diagnosis to Device Level*. Proc. SERT Microtest, April 1979, pp. 92–100.

(vii) Instruction-code checkout

141. V.P. Srini. *Fault Diagnosis of Microprocessor Systems*. IEEE Computer, Vol. 10, No. 1, Jan. 1977, pp. 60–65.

142. D. Peckett. *Fault-finding with Aid of Self-test Programs*. Practical Computing, Vol. 2, No. 12, Dec. 1979, pp. 102–107.

10.2.8 Chapter 9

(i) Testability

143. J.E. Stephenson and J. Grason. *A Testability Measure for Register-Transfer-Level Digital Circuits*. Proc. 6th FTCS, 1976, pp. 156–161.

144. W.J. Dejka. *Measure of Testability in Device and System Design*. Proc. 20th Midwest Symposium on Circuits and Systems, 1977, pp. 39–52.

145. M.A. Breuer. *New Concepts in Automated Testing of Digital Circuits*. Proc. EEC Symp. on CAD of Digital Electronic Circuits and Systems, Brussels, 1978, pp. 69–92.

146. J.A. Dussault. *A Testability Measure*. Proc. STC, 1978, pp. 113–116.

147. J. Grason. *TMEAS–a Testability Measurement Program*. Proc. 16th DA, 1979, pp. 156–161.

148. C.T. Wood. *A Quantitative Measure of Testability*. Proc. IEEE AutoTest Conference, 1979, pp. 286–291.

149. F. Danner and W. Consolla. *An Objective PCB Testability Rating System*. Proc. TC, 1979, pp. 23–28.

150. L.H. Goldstein. *Controllability/Observability Analysis for Digital Circuits*. IEEE Trans. Circuits and Systems, Vol. CAS-26, 1979, pp. 685–693.

151. P.G. Kovijanic. *Computer Aided Testability Analysis*. Proc. IEEE AutoTest Conference, 1979, pp. 292–294.

152. R.G. Bennetts, C.M. Maunder, G.D. Robinson. *Computer-aided Measurement of Logic Testability–the CAMELOT Program*. Proc. IEEE International Conference on Circuits and Computers, Oct. 1980, pp. 1162–1165.

(ii) Scan-In, Scan-Out (excluding LSSD: see following section)

153. W.C. Carter, H.C. Montgomery, R.J. Preiss, and H.J. Reinheimer. *Design of Serviceability Features for the IBM System 360*. IBM Journal Research and Development, Vol. 8, April 1964, pp. 115–126.

154. M.J.Y. Williams and J.B. Angell. *Enhancing Testability of LSI Circuits via Test Points and Additional Logic*. IEEETC, Vol. C-22, 1973, pp. 46–60.

155. A. Toth and C. Holt. *Automated Data-base-driven Digital Testing*. IEEE Computer, Vol. 7, No. 1, Jan. 1974, pp. 13–19.

156. J.H. Stewart. *Future Testing of Large LSI Circuit Cards*. Proc. STC 1977, pp. 6–15.

157. A. Yamada et al. *Automatic Test Generation for Large Digital Circuits*. Proc. 14th DA, 1977, pp. 78–83.

158. A. Yamada et al. *Automatic System Level Test Generation and Fault Location for Large Digital Systems.* Proc. 15th DA., 1978, pp. 347–352.

159. S. Funatsu, N. Wakatsuki, and A. Yamada. *Designing Digital Circuits with Easily Testable Considerations.* Proc. STC, 1978, pp. 98–102.

160. S. Funatsu, N. Wakatsuki, and A. Yamada. *Easily Testable Design of Large Digital Circuits.* NEC Journal Research and Development, No. 54, 1979, pp. 49–55.

(iii) Level-Sensitive Scan Design

161. E.I. Muehldorf. *Designing LSI Logic for Testability.* Proc. STC, 1976, pp. 45–49.

162. T.W. Williams and E.B. Eichelberger. *Random Patterns within a Structured Sequential Logic Design.* Proc. STC, 1977, pp. 19–26.

163. E.B. Eichelberger and T.W. Williams. *A Logic Design Structure for Testability.* Proc. 14th DA, 1977, pp. 463–468.

164. P. Bottorf and E.I. Muehldorf. *Impact of LSI on Complex Digital Circuit Board Testing.* Proc. IEEE Electro 77 Conference, 1977, pp. 1–12.

165. G.H. Stange. *A Test Methodology for Large Logic Networks.* Proc. 15th DA, 1978, pp. 103–109.

166. E.B. Eichelberger and T.W. Williams. *A Logic Design Structure for LSI Testability.* Journal DAFTC, Vol. 2, 1978, pp. 165–178.

167. H.E. Jones and R.F. Schauer. *An Approach to a Testing System for LSI.* Proc. EEC Symposium on CAD of Digital Electronics Circuits and Systems, Brussels, 1978, pp. 257–274.

168. L.A. Stolte and N.C. Berglund. *Design for Testability of the IBM System 38.* Proc. TC, 1979, pp. 29–36.

169. N.C. Berglund. *Level-sensitive Scan Design Tests Chips, Boards, Systems.* Electronics International, Vol. 52, No. 6, March 15, 1979, pp. 108–110.

170. T.J. Frechette and F. Tanner. *Support Processor Analyses Errors Caught by Latches.* Electronics International, Vol. 52, No. 23, Nov. 8, 1979, pp. 116–118.

(iv) Signature Analysis

171. G. Gordon and H. Nadig. *Hexadecimal Signatures Identify Trouble Spots in Microprocessor Systems.* Electronics International, Vol. 50, No. 5, March 3, 1977, pp. 89–96.

172. Hewlett-Packard. *A Designers Guide to Signature Analysis.* Application Note 222, April 1977.

173. Hewlett-Packard. *Implementing Signature Analysis for Production Testing.* Application Note 222-1, 1977.

174. L. Badagliacca and R. Catterton. *Combining Diagnosis and Emulation Yields Fast Fault-Finding.* Electronics International, Vol. 50, No. 23, Nov. 10, 1977, pp. 107–110.

175. H. Nadig. *Testing a Microprocessor Product using Signature Analysis.* Proc. STC, 1978, pp. 159–169.

176. M. Neil and R. Goodner. *Designing a Serviceman's Needs into Microprocessor-based Systems.* Electronics International, Vol. 52, No. 5, March 1, 1979, pp. 122–128.

177. C. Pynn. *In-circuit Tester Using Signature Analysis Adds Digital LSI to Its Range.* Electronics International, Vol. 52, No. 11, May 24, 1979, pp. 153–157.

178. D.G. West. *In-circuit Emulation and Signature Analysis: Vehicles for Testing Microprocessor-based Products.* Proc. ST, 1979, pp. 345–353.

179. J. Humphrey and K. Firooz. *ATE Brings Speedy, Complete Testing via Signature Analysis to LSI-board Production.* Electronic Design, Vol. 28, No. 3, Feb. 1, 1980, pp. 75–79.

(v) Testable logic: practical guidelines and surveys

180. C.V. Ramamoorthy. *A Structural Theory of Machine Diagnosis.* Proc. AFIPS Spring Joint Computer Conference, Vol. 30, 1967, pp. 743–756.

181. A.D. Friedman. *Fault Detection in Redundant Circuits.* IEEETEC, Vol. EC-16, 1967, pp. 99–100.

182. F.R. Boswell. *Designing Testability Into Complex Logic Boards.* Electronics International, Vol. 45, No. 17, Aug. 14, 1972, pp. 116–119.

183. D. Schneider. *Designing Logic Boards for Automatic Testing*. Electronics International, Vol. 47, No. 15, July 25, 1974, pp. 100–104.

184. J.P. Hayes and A.D. Friedman. *Test Point Placement to Simplify Fault Detection*. IEEETC, Vol. C-23, 1974, pp. 727–735.

185. R. Dandapani and S.M. Reddy. *On the Design of Logic Networks with Redundancy and Testability Considerations*. IEEETC, Vol. C-23, 1974, pp. 1139–1149.

186. P.L. Writer. *Design for Testability*. Proc. IEEE Automated Support Systems Conference, 1975, pp. 84–87.

187. R.G. Bennetts and R.V. Scott. *Recent Developments in the Theory and Practice of Testable Logic Design*. IEEE Computer, Vol. 9, No. 6, June 1976, pp. 47–63.

188. D. Tose. *Digital Logic Board Design with Test Needs in Mind*. Electronic Engineering, Vol. 48, No. 586, Dec. 1976, pp. 73–75 (Part 1); Vol. 49, No. 587, Jan. 1977, pp. 46–48 (Part 2).

189. J. Mittelbach. *Put Testability into PC Boards*. Electronic Design, Vol. 26, No. 12, June 7th, 1978, pp. 128–131.

190. G. Foley. *Designing Microprocessor Boards for Testability*. Proc. STC, 1978, pp. 176–179.

191. J. Boney and E. Rupp. *Let Your Next Microprocessor Check Itself and Cut Down Your Testing Overhead*. Electronics Design, Vol. 27, No. 18, Sept. 1, 1979, pp. 100–105.

192. T.W. Williams and K.P. Parker. *Testing Logic Circuits and Designing for Testability*. IEEE Computer, Vol. 12, No. 10, Oct. 1979, pp. 9–21.

193. M.D. Lippman and E.S. Donn. *Design Forethought Promotes Easier Testing of Microcomputer Boards*. Electronics International, Vol. 52, No. 2, Jan. 18, 1979, pp. 113–119.

194. J.P. Hayes and E.J. McCluskey. *Testability Considerations in Microprocessor-based Designs*. IEEE Computer, Vol. 13, No. 3, March 1980, pp. 17–26.

195. Hewlett-Packard. *Designing Digital Circuits for Testability*. Application Note 210-4.

196. GenRad. *How to Design Logic Boards for Easier Automatic Testing and Troubleshooting*. Technical publication.

197. Computer Automation. *Design for Testability*. Technical publication (AN-104-378-500).

Appendix

The D-Algorithm

The concept of generating a test for a specified logical fault by sensitizing a path from the site of the fault to an observable output node is well known. There have been many attempts to formalize this procedure in such a way as to make it suitable for implementation on a digital computer. This appendix describes one such attempt, the D-algorithm, developed originally by J. P. Roth of IBM. The appendix is designed to be relatively independent of the rest of the book.

In its pure form, the D-algorithm has certain practical disadvantages and is limited to circuits that do not contain stored-state devices (such as flip-flops) or global feedback (sequential circuits with both feedforward and feedback). Nevertheless, the algorithm is important inasmuch as it does contain the mechanisms for handling the many-choice situations that arise in test-pattern generation as well as problems arising from fault-effect reconvergence. Such problems can either be caused by structural reconvergence (i.e., a fault-effect that fans out and propagates along multiple paths which then reconverge onto a common device) or by temporal reconvergence (as could occur in circuits with global feedback). Potentially, therefore, the D-algorithm could be extended to cover circuits with global feedback, provided some additional timing parameter could be included into the basic algorithm. The appendix concludes with comments on the extension of the D-algorithm.

A1.1 *D-ALGORITHM*

Before discussing the details of the D-algorithm, let us recapitulate the basic processes of the path sensitization technique. First, we must have some knowledge of the circuit topology and the fault-free behavior of each element within the circuit. Second, we must postulate a source fault and attempt to drive the effect of this fault to some observable output position (primary output or test point) by chaining local sensitive paths together. This process specifies logic levels on certain other connections within the circuit. The final stage is to consolidate the test by backtracking to the primary inputs, checking that the levels can in fact be established. In some cases this is not possible for the particular path chosen, and when this occurs the process must be repeated for another choice of path, single or multiple.*

These processes require the following:

(i) a knowledge of the normal fault-free behavior of each element within the circuit,

(ii) a way of indicating the status, sensitive or otherwise, of each connection involved in creating a sensitive path,

(iii) a knowledge of the fault-transmission and fault-blocking properties of each element within the circuit,

(iv) a knowledge of the fault-generation properties of each element within the circuit,

(v) some mechanism for keeping track of the particular path choices that have been studied so far, and

(vi) some procedure for evaluating the positive or negative nature of fault-effect reconvergence if it occurs.

As far as points (i)–(iv) are concerned, the D-algorithm covers these by introducing the notions of a singular cover (point (i)); the symbols D and $\overline{\text{D}}$ (point (ii)); propagating and non-propagating D-cubes (point (iii)); and D-cubes of failure (point (iv)). Each of these notions are now discussed in detail and illustrated by application to a 2-input NAND gate. Subsequently, the singular cover, propagation D-cubes, non-

---

*In a circuit with fanout, it is possible for a source fault to propagate simultaneously along more than one path. If this happens, the path sensitization is called multiple.

propagation D-cubes, and D-cubes of failure for all the basic gates are presented before elaborating on points (v) and (vi). In all cases the results are derived in an informal manner and the proofs are seen "by inspection." Because of the importance of the D-algorithm and the fact that it can be extended to more complex logic circuit elements and to other types of fault model, the formal derivations of propagation D-cubes and D-cubes of failure are also given. The derivation of the singular cover is discussed in the next section.

A1.1.1. Singular cover and cube notation

The *singular cover* of a logic gate is derived from its truth table and is a more compact form for expressing the fault-free behavior. It is actually a list of the prime implicants that cause the gate output to equal 1 (sometimes called the *ON-array*) and those that cause the gate output to equal 0 (the *OFF-array*). The description makes use of the functional values 0, 1, and X, where X stands for "don't care" and can take either the 0 or 1 value. Fig. A1.1 shows the truth table and corresponding singular cover for the 2-input NAND gate. The singular cover summarizes the fact that if any single input is 0, the output must be 1, whereas it is necessary for all inputs to be 1 for the output to become 0.

The other term we must introduce and define is a *cube,* sometimes called an *n-tuple.**

A cube is an ordered set of n *symbols* or coordinates such that:

(i) each symbol position identifies with a particular connection carrying a logic value, i.e., primary input, internal connection, or primary output, and

(ii) the value of the symbol identifies the status of the connection.

In the example of Fig. A1.1, the singular cover is said to consist of three *singular cubes* defined by (0X1), (X01), and (110). In these cubes, the left-hand position identifies with the variables c_1, the middle position with the variable c_2, and the right-hand with c_3. The value of the symbols in each cube is constrained to be 0, 1, or X. In the following section, the

---

*Both terms are used in the literature, but cube is more common and is preferred here.

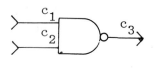

c_1	c_2	c_3
0	0	1
0	1	1
1	0	1
1	1	0

c_1	c_2	c_3
0	X	1
X	0	1
1	1	0

Truth-table   Singular cover

FIGURE A1.1 Nand Gate: Singular Cover

D and $\bar{\text{D}}$ symbols will be added as further possible values, and the cube will become known as a *D-cube*.

Because of the positional importance of each symbol in a cube, it is necessary to include the connection number as a subscript. (The *terminal numbering convention* is recommended here. In this convention, the numbering of connections starts at the primary inputs and is ordered such that each logic device output is assigned an integer greater than any integer used for the inputs to the device. This also applies to fanout points.) This means that the three singular cubes should be written $(0_1 X_2 1_3)$, $(X_1 0_2 1_3)$, $(1_1 1_2 0_3)$. For cubes written in the text, this notation is adopted. For cubes listed in figures or tables, where the positional relationship is included at the head, the subscript is omitted.

Finally, for those interested in the reasons for the word "cube," the name derives from the fact that any n-variable Boolean function can be mapped onto an n-dimensional space. (An n-variable K-map is a 2-dimensional version of this n-variable space.) In general the collection of 2^n possible points resulting from the various combinations of n variables is said to form the *vertices* of an *n-cube,* and this form of representation is useful to those concerned with the development of switching theory. The *cubical* representation of a function therefore consists of the set of vertices corresponding to the miniterms of the function. Single vertices are known as 0-cubes. If two vertices are adjacent, they may be represented either by the two 0-cubes or by a single cube in which one of the variables is represented by X. This is known as a 1-cube and corresponds to the combination of two adjacent terms on a K-map. Likewise, 2-cubes, 3-cubes, . . . , k-cubes are similarly defined.

In general, therefore, a Boolean function can be expressed as the Boolean sum of a number of k-cubes where each 1-cube defines a vertex, a line, a plane, etc., in the n-cube.

In the D-algorithm, the word "cube" has been adopted to define any ordered array of symbols.

A1.1.2 The D symbol and propagating D-cube

So far we have used only the symbols 0, 1 and X to define the input/output behavior of a logic element. We now introduce the D symbol as a special symbol to be used to indicate sensitivity status on element inputs and outputs. The D-symbol is constrained to take the values 0 or 1 and we may also make use of the complemented value $\bar{D}$ if necessary.

The use of the D-symbol is explained with reference to the 2-input NAND gate in Fig. A1.1. We know that a local path exists from c_1 to c_3 if c_2 is held at 1. In other words, a fault-effect transmitted to c_1 from some predecessor element will continue to be propagated through to c_3 provided c_2 is held at 1. Furthermore, if the change in c_1 is from $0 \rightarrow 1$, the corresponding change in c_3 is from $1 \rightarrow 0$. Alternatively, if the change in c_1 is from $1 \rightarrow 0$, c_3 changes from $0 \rightarrow 1$. This information is part of the fault-transmission property of the NAND gate and is summarized by the *propagating D-cube** $(D_1 1_2 \bar{D}_3)$.

In this type of cube, the symbol D has been used to indicate the sensitivity of the connections c_1 and c_3. The value that D can take is either 0 or 1 but whichever one it takes, the value of $\bar{D}$ must be the complement. This means that the propagating D-cube $(D_1 1_2 \bar{D}_3)$ represents the two cubes $(0_1 1_2 1_3)$ and $(1_1 1_2 0_3)$ corresponding to the assignments $D = 0$, $\bar{D} = 1$ and $D = 1$, $\bar{D} = 0$ respectively. Because of the dual assignment possibility, we can define the *dual* of a propagating D-cube simply by changing all D's to $\bar{D}$'s and $\bar{D}$'s to D, i.e., if $(D_1 1_2 \bar{D}_3)$ is a propagating D-cube, then so is $(\bar{D}_1 1_2 D_3)$.

Other propagating D-cubes exist for the 2-input NAND gate, namely $(1_1 D_2 \bar{D}_3)$ and $(D_1 D_2 \bar{D}_3)$. The first of these is similar to the $(D_1 1_2 \bar{D}_3)$ cube, merely reversing the roles of the connections c_1 and c_2. The $(D_1 1_2 \bar{D}_3)$ and $(1_1 D_2 \bar{D}_3)$ D-cubes are both examples of *single-input*

*Roth used the term "primitive" rather than "propagating." Other workers have preferred the term "propagating" and it is the preferred term here.

propagating D-cubes. This means that only one of the input coordinates has a D or $\bar{D}$ status. The $(D_1 D_2 \bar{D}_3)$ cube is an example of a *multi-input propagating D-cube* and expresses the fact that it is necessary for, in this case, two inputs to change simultaneously before the output changes. This would correspond either to a multiple fault situation with fault-effects arriving from some predecessor elements or to a positive reconvergence situation for a single source fault.

We can also define *non-propagating D-cubes:* $(0_1 D_2 1_3)$, $(D_1 0_2 1_3)$, and $(D_1 \bar{D}_2 1_3)$ would be non-propagating D-cubes for the 2-input NAND gate. The main feature here is that the value of the output coordinate is *not* dependent on changes in the value of the input coordinate. This would cause termination of a sensitive path, and the $(D_1 \bar{D}_2 1_3)$ cube in particular again relates either to a multiple-fault situation or to a negative reconvergence situation for a single-source fault.

In summary therefore, the main points are as follows:

1. Propagating D-cubes summarize the fault-transmission properties of an element and identify the sensitive input(s) and output by use of the symbol D and its complement $\bar{D}$.
2. The symbol D can take either the value 1 or 0 but whichever one it takes, $\bar{D}$ always takes the complement.
3. The dual of a propagating D-cube is also a propagating D-cube and is formed by interchanging D's for $\bar{D}$ and $\bar{D}$'s for D.
4. It is also possible to define non-propagating D-cubes. These cubes summarize the non-propagating properties of the element. The dual of a non-propagating D-cube is also a non-propagating D-cube and is formed as in 3 above.

Fig. A1.2 contains the complete list of propagating and non-propagating D-cubes for the 2-input NAND gate. (The formal procedure for deriving these D-cubes from the singular cover is described in section A1.2.)

A1.1.3 Failure D-cubes

So far, the D-cubes we have defined relate to the fault-transmission properties of an element rather than to the fault-generation properties.

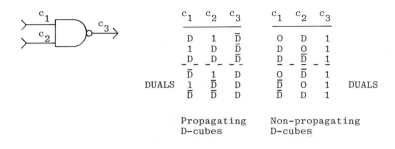

c_1	c_2	c_3		c_1	c_2	c_3	
D	1	$\bar{D}$		0	D	1	
1	D	$\bar{D}$		D	0	1	
D	D	$\bar{D}$		D	$\bar{D}$	1	
$\bar{D}$	1	D		0	$\bar{D}$	1	
DUALS 1	$\bar{D}$	D		$\bar{D}$	0	1	DUALS
$\bar{D}$	$\bar{D}$	D		$\bar{D}$	D	1	

Propagating D-cubes	Non-propagating D-cubes

FIGURE A1.2 Nand Gate: Propagating and Non-Propagating D-Cubes

We now define a *failure D-cube* as being a cube that carries fault-generation information and, because of this fact, the interpretation of the symbol D must be according to whether the element is operating in a fault-free or faulty mode.

Failure D-cubes are constructed from truth tables relating to the fault-free and faulty behavior of the element and can be made to relate to any type of failure mechanism that modifies the logical function of the element. (The general procedure for doing this is described in section A1.2. We will restrict our attention to single s-a-1, s-a-0 failures occurring on element inputs or outputs only, and the validity of the cubes will be seen "by inspection.")

Fig. A1.3 shows the failure D-cubes for the single, s-a-1, and s-a-0 faults on the 2-input NAND gate.

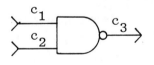

c_1	c_2	c_3	Fault-cover
0	0	D	$c_3/0$
0	1	D	$c_1/1, c_3/0$
1	0	D	$c_2/1, c_3/0$
1	1	$\bar{D}$	$c_1/0, c_2/0, c_3/1$

Failure D-cubes

FIGURE A1.3 Nand Gate: Failure D-Cubes

The interpretation of the D in any failure D-cube is that the assignment $D = 1$ ($\overline{D} = 0$) is taken to correspond to the normal fault-free value whereas the assignment $D = 0$ ($\overline{D} = 1$) corresponds to the behavior of the element in the presence of the fault.* The failure D-cube $(0_1 1_2 D_3)$ therefore suggests that under the fault-free behavior, the input values $c_1 = 0$, $c_2 = 1$ produces an output value $c_3 = 1$. If c_1 is s-a-1 or c_3 is s-a-0, however, the value will be $c_3 = 0$. Similar interpretations can be made about the other two failure D-cubes.

Note:

> (i) because the value assigned to a D in a failure D-cube is constrained, the dual of a failure D-cube has no meaning, and
> (ii) failure D-cubes for a primary input coming in on a connection c_i that subsequently fans out is written as D_i (input s-a-0) or $\overline{D}_i$ (input s-a-1).

A1.1.4 Intersection of D-cubes

Armed with these mechanisms for describing the fault-transmission and fault-generation properties of the logic elements that make up the circuit, we are now almost ready to develop the full D-algorithm. The three basic processes for test-pattern generation are:

1. Postulate the fault, i.e., identify a failure D-cube covering the fault.
2. Choose a path, or set of paths, and attempt to drive the effect of the fault to a primary output, i.e., drive the "D" forward using the propagating D-cubes. (This is called the *D-drive* stage of the D-algorithm.)
3. Check the consistency of the path using the fault-free singular cover information. If the consistency fails, select another possible path and try again. If the consistency is successful, a test has been found.

Fundamental to the D-drive and consistency operations above is the process of intersecting D-cubes. This is illustrated by applying the D-algorithm to the fault c_1 s-a-0 in the circuit shown in Fig. A1.4(a).

*Although the assignment can be arbitrary, this particular convention is the accepted one.

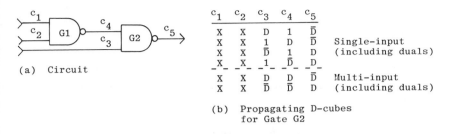

(a) Circuit

c_1	c_2	c_3	c_4	c_5	
X	X	D	1	$\overline{D}$	
X	X	1	D	$\overline{D}$	Single-input
X	X	$\overline{D}$	1	D	(including duals)
X	X	1	$\overline{D}$	D	
X	X	D	D	$\overline{D}$	Multi-input
X	X	$\overline{D}$	$\overline{D}$	D	(including duals)

(b) Propagating D-cubes for Gate G2

FIGURE A1.4 D-Cube Intersection

The failure D-cube for c_1 s-a-0 is given in Fig. A1.3:

$$d_1 = (1_1 1_2 X_3 \overline{D}_4 X_5)$$

Note that spare positions for the connections c_3 and c_5 have been included in the cube but because we are only concerned with gate G1 at the moment, the assignments have been left at X.

The next stage is to try and match the entries in d_1 with one of the propagating D-cubes for gate G2. These are shown in Fig. A1.4(b) and, in this case, include X's on the connections c_1 and c_2. Also, because we are only interested in a single fault situation and there is no possibility of fault reconvergence, we can only consider the single-input propagating D-cubes.

In trying to match d_1 with the single-input propagating D-cubes, the main problem is matching the $\overline{D}_4$ coordinate. The only propagating D-cube that matches is $(X_1 X_2 1_3 \overline{D}_4 D_5)$, and this modifies d_1 to produce a composite result d_2:

$$d_2 = (1_1 1_2 1_3 \overline{D}_4 D_5)$$

This completes the D-drive process: d_2 now contains a D-status value on the connection c_5 and, for this example, the algorithm terminates. The test is defined by d_2 and specifies $c_1 = c_2 = c_3 = 1$ with fault-free values of 0 on c_4 and 1 on c_5 corresponding to the $D = 1$, $\overline{D} = 0$ convention for fault-free behavior. The alternative assignment to D gives the values on c_4 and c_5 if the fault c_1 s-a-0 is present.

The D-cube d_2 above was obtained in an intuitive manner by comparing d_1 with all the single-input propagating D-cubes for gate G2, looking for a match on the symbol $\overline{D}_4$. In doing this comparison we rejected the first three cubes in Fig. A1.4(b) because the value on c_4 was either 1 or D but not $\overline{D}$. This process of comparing two D-cubes looking for a composite result is called *intersection* and can be defined precisely in the following way.

Let α and β be two cubes of length n such that:

$$\alpha = (\alpha_1, \alpha_2, \ldots, \alpha_i, \ldots, \alpha_n)$$
$$\beta = (\beta_1, \beta_2, \ldots, \beta_i, \ldots, \beta_n)$$
and $\alpha_i, \beta_i = 0, 1, X, D$ or $\overline{D}$ for all i.

The intersection between α and β is written $\alpha \cap \beta$ and is defined by:

$$\alpha \quad \beta = ((\alpha_1 \cap \beta_1), (\alpha_2 \cap \beta_2), \ldots, (\alpha_i \cap \beta_i), \ldots, (\alpha_n \cap \beta_n)$$

where, for each $\alpha_i \beta_i$ pair

$$\alpha_i \cap \beta_i = \alpha_i \text{ if } \alpha_i = \beta_i$$
$$= \alpha_i (\beta_i) \text{ if } \beta_i (\alpha_i) = X$$
$$= \emptyset, \text{ the null result otherwise.}$$

These results are summarized in Table A1.1.

β_i \ α_i	0	1	X	D	$\overline{D}$
0	0	$\emptyset$	0	$\emptyset$	$\emptyset$
1	$\emptyset$	1	1	$\emptyset$	$\emptyset$
X	0	1	X	D	$\overline{D}$
D	$\emptyset$	$\emptyset$	D	D	$\emptyset$
$\overline{D}$	$\emptyset$	$\emptyset$	$\overline{D}$	$\emptyset$	$\overline{D}$

$\alpha_i \cap \beta_i$

TABLE A1.1 Rules for Intersecting D-Cube Coordinates

Finally, $\alpha \cap \beta = \emptyset$, an empty cube, if any $\alpha_i \cap \beta_i = \emptyset$.

As examples of the intersection process, consider the D-cubes

$$\alpha = (1\ 1\ X\ \bar{D}\ X)$$
$$\beta = (X\ X\ 1\ \bar{D}\ D)$$
$$\sigma = (X\ X\ D\ 1\ D)$$

α corresponds to the failure D-cube d_1 for gate G1 and β and σ are two of the propagating D-cubes for gate G2.

Applying the rules of intersection, we obtain:

$$\alpha \cap \beta = ((1 \cap x), (1 \cap x), (x \cap 1), (\bar{D} \cap \bar{D}), (x \cap D))$$
$$= (1\ 1\ 1\ \bar{D}\ D), \text{ i.e., } d_2$$

whereas:

$$\alpha \cap \sigma = ((1 \cap X), (1 \cap X), (X \cap D), (\bar{D} \cap 1), (X \cap D))$$
$$= (1\ 1\ D\ \emptyset\ D)$$
$$= \emptyset$$

A1.1.5 Application to an example

We are now in a position to demonstrate the full D-algorithm and we will make use of the example circuit shown in Fig. A1.5.

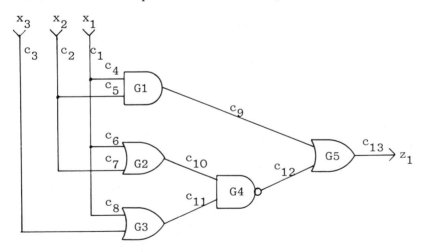

FIGURE A1.5 Example Circuit

For reference purposes, Table A1.2 lists the singular cover, propagating D-cubes, non-propagating D-cubes and failure D-cubes for 2-input AND, OR, NAND, NOR, and EOR gates. The table also includes the INVERT gate and a fanout point. The latter may be considered to be either a single-input, multi-output element or a collection of single-input, single-output elements.

In general, however, the following conditions are true for a fanout point:

Singular cover: a 1 (0) on the trunk will produce 1's (0's) on all branches.

Propagating D-cubes: a D ($\overline{D}$) on the trunk will propagate a D ($\overline{D}$) to all branches.

Non-propagating D-cubes: it is not possible to prevent the propagation of a trunk D ($\overline{D}$) to any of the branches.

Failure D-cubes: these exist for stuck-at faults on the trunk inasmuch as a 1 (0) on the trunk will create a failure D-cube with $\overline{D}$'s (D's) on all branches. Unfortunately, it is not necessarily true that a fault on a single branch will affect either the trunk or any other branch. Consider a branch input to a TTL gate as being:

(a) open-circuit, or
(b) bridged to the +5v rail

In both cases, the TTL input appears to be s-a-1 but the effect on other connections is quite different.

In general, therefore, failure D-cubes for branch faults will depend both on the exact nature of the failure mechanism and on the technology. Table A1.2 shows only the failure D-cubes relating to the trunk faults. The validity of all the results in this table can either be seen "by inspection" or proved using the formal procedure described later (in section A1.2). Using this table, the singular cover for the whole circuit can be derived. This is shown in Table A1.3.

In this table, blank entries denote unspecified values and can be interpreted as having a don't care value X. Also, because the circuit contains 13 separately-identified connections, the length of any cube, propagation, or failure should be correspondingly 13. For brevity (and clarity), however, only those coordinates that have specified 0, 1, D, or $\overline{D}$ values are defined.

GATE	SINGULAR COVER 1	2	3	PROPAGATING D-CUBES* 1	2	3	NON-PROPAGATING D-CUBES* 1	2	3	FAILURE D-CUBES AND FAULT-COVER** 1	2	3	
AND	0	X	0	1	D	D	0	D	0	1	0	D̄	2/1,3/1
	X	0	0	D	1	D	D	0	0	0	1	D̄	1/1,3/1
	1	1	1	1	D̄	D̄	0	D̄	0	1	1	D	1/0,2/0,3/0
				D̄	1	D̄	D̄	0	0	0	0	D̄	3/1
				D	D	D	D	D̄	0				
				D̄	D̄	D̄	D̄	D	0				
OR	1	X	1	0	D	D	1	D	1	1	0	D	1/0,3/0
	X	1	1	D	0	D	D	1	1	0	1	D	2/0,3/0
	0	0	0	0	D̄	D̄	1	D̄	1	0	0	D̄	1/1,2/1,3/1
				D̄	0	D̄	D̄	1	1	1	1	D	3/0
				D	D	D	D	D̄	1				
				D̄	D̄	D̄	D̄	D	1				
NAND	0	X	1	1	D	D̄	0	D	1	1	0	D	2/1,3/0
	X	0	1	D	1	D̄	D	0	1	0	1	D	1/1,3/0
	1	1	0	1	D̄	D	0	D̄	1	1	1	D̄	1/0,2/0,3/1
				D̄	1	D	D̄	0	1	0	0	D	3/0
				D	D	D̄	D	D̄	1				
				D̄	D̄	D	D̄	D	1				
NOR	1	X	0	0	D	D̄	1	D	0	1	0	D̄	1/0,3/1
	X	1	0	D	0	D̄	D	1	0	0	1	D̄	2/0,3/1
	0	0	1	0	D̄	D	1	D̄	0	0	0	D	1/1,2/1,3/0
				D̄	0	D	D̄	1	0	1	1	D̄	3/1
				D	D	D̄	D	D̄	0				
				D̄	D̄	D	D̄	D	0				
EOR	0	0	0	0	D	D	D	D	0	0	0	D	1/1,2/1,3/1
	0	1	1	D	0	D	D̄	D̄	0	0	1	D	1/1,2/0,3/0
	1	0	1	0	D̄	D̄	D	D̄	1	1	0	D	1/0,2/1,3/0
	1	1	0	D̄	0	D̄	D̄	D	1	1	1	D̄	1/0,2/0,3/1
				1	D	D̄							
				D	1	D̄							
				1	D̄	D							
				D̄	1	D							
INV	1	-	0	D	-	D̄	None			0	-	D	1/1,3/0
	0	-	1	D̄	-	D				1	-	D̄	1/0,3/1
FAN-OUT	1	1	1	D	D	D	None			0	D̄	D̄	1/1
	0	0	0	D̄	D̄	D̄				1	D	D	1/0

* Includes all single-input and multi-input cubes plus duals

** Failure D-cubes for single s-a-1, s-a-0 faults on inputs or outputs

TABLE A1.2 D-Algorithm: Properties of Logic Elements

Application 1. Test for c_1 s-a-1

The final objective of the D-algorithm is to derive a cube defining test input conditions for the postulated source fault as defined by the initial failure D-cube. This cube is called a *test cube,* abbreviated to tc, and the initial version of the test cube, tc_0, is defined by the failure D-cube covering the postulated fault. For this particular fault, there is a choice of failure D-cubes depending upon what path or set of paths we wish to

ELEMENT	c_1	c_2	c_3	c_4	c_5	c_6	c_7	c_8	c_9	c_{10}	c_{11}	c_{12}	c_{13}
c_1 Fan-out	O			O		O		O					
	1			1		1		1					
c_2 Fan-out		O			O		O						
		1			1		1						
G1				O	X				O				
				X	O				O				
				1	1				1				
G2						1	X		1				
						X	1		1				
						O	O		O				
G3				X					1		1		
				1					X		1		
				O					O		O		
G4										O	X	1	
										X	O	1	
										1	1	O	
G5										1		X	1
										X		1	1
										O		O	O

TABLE A1.3 Singular cover for Example Circuit

attempt a D-drive. As a first choice we will attempt to drive the D through all possible gates, i.e., G1, G2, and G3, and this will define tc_0 to be:

$$tc_0 = (0_1 \; \bar{D}_4 \; \bar{D}_6 \; \bar{D}_8)$$

Associated with any test cube is a list of gates that have at least one input coordinate, specified as a D or $\bar{D}$, and whose output coordinate has not yet been specified. These are the gates to which a D or $\bar{D}$ has been driven, but the D propagating (or non-propagating) conditions have not yet been determined. (This list is called the *D-frontier* or *D-fanout*, abbreviated to Df.) A Df has the same subscript as the associated tc, and Df_0 is given by:

$$Df_0 = (G1, G2, G3)$$

The next stage is to intersect tc_0 with the propagating D-cubes of each of the gates listed in Df_0 until either all intersections are completed successfully or the test cube is reduced to the empty cube ϕ. The latter would indicate no test condition along the path set chosen.

These stages are detailed below, starting with gate G1. Note that we take one gate at a time, modifying the test cube and D-frontier accordingly. We must also select only from the single-input propagation D-cubes unless D-reconvergence occurs (see below).

Gate G_1.
$$tc_1 = tc_0 \cap (\bar{D}_4 1_5 \bar{D}_9)$$
$$= (0_1 \bar{D}_4 1_5 \bar{D}_6 \bar{D}_8 \bar{D}_9)$$
$$Df_1 = (G_2, G_3, G_5)$$

Gate G_2.
$$tc_2 = tc_1 \cap (\bar{D}_6 0_7 \bar{D}_{10})$$
$$= (0_1 \bar{D}_4 1_5 \bar{D}_6 0_7 \bar{D}_8 \bar{D}_9 \bar{D}_{10})$$
$$Df_2 = (G_3, G_4, G_5)$$

Gate G_3.
$$tc_3 = tc_2 \cap (0_3 \bar{D}_8 D_{11})$$
$$= (0_1 0_3 \bar{D}_4 1_5 \bar{D}_6 0_7 \bar{D}_8 \bar{D}_9 \bar{D}_{10} \bar{D}_{11})$$
$$Df_3 = (G_4 (2), G_5)$$

Comment: We have now succeeded in driving the D to two different input coordinates of G_4. This is *D-reconvergence* and is denoted by $G_4(2)$ in Df_3. It means that any further propagation through G_4 must be based on a multi-input propagation D-cube, in this case a 2-input cube.

Gate G_4.
$$tc_4 = tc_3 \cap (\bar{D}_{10} \bar{D}_{11} D_{12})$$
$$= (0_1 0_3 \bar{D}_4 1_5 \bar{D}_6 0_7 \bar{D}_8 \bar{D}_9 \bar{D}_{10} \bar{D}_{11} D_{12})$$
$$Df_4 = (G_5 (2)$$

Comment: Because the intersection was successful, we have identified positive reconvergence at gate G_4.

Gate G_5.
$$tc_5 = tc_4 \cap (D_9 D_{12} D_{13}) = \emptyset, \text{ try again}$$
$$= tc_4 \cap (\bar{D}_9 \bar{D}_{13} \bar{D}_{13}) = \emptyset, \text{ still failed}$$

Comment: The failure to drive the D's through gate G_5 leads to the conclusion that a test for this fault that simultaneously sensitizes all paths from the site of the faults to the primary output does not exist. Furthermore, by noting that tc_4 does intersect successfully with the 2-input non-propagating D-cube for gate G_5 to give

$$tc_5 = tc_4 \cap (\bar{D}_9 D_{12} 1_{13})$$
$$= (0_1 0_3 \bar{D}_4 1_5 \bar{D}_6 0_7 \bar{D}_8 \bar{D}_9 \bar{D}_{10} \bar{D}_{11} D_{12} 1_{13})$$

we conclude that the reason for failing to compute a test is that negative reconvergence has occurred at gate G_5.

Having failed to generate a test for c_1 s-a-1 by propagating the effect along all paths simultaneously, the D-algorithm can be applied again, but this time to a subset of paths. As a second choice, therefore, we will try driving the D's through gates G_2 and G_3 only. A necessary condition for this to take place is that all other possible paths must be blocked. This is achieved by intersecting the test cube with the non-propagating D-cubes of those gates that are not to be part of the sensitive path.

As before, the initial failure D-cube and D-frontier is given by:

$$tc_0 = (0_1 \bar{D}_4 \bar{D}_6 \bar{D}_8)$$
$$Df_0 = (\bar{G}_1, G_2, G_3)$$

Note that in the D-frontier, G_1 is denoted by $\bar{G}_1$, the overline (bar) denoting that non-propagating D-cubes must be used when this gate is considered, rather than the propagating D-cubes. Apart from this point, the procedure is identical to the previous example and is as follows:

Gate G_1. $\quad tc_1 = tc_0 \cap (\bar{D}_4 0_5 0_9)$
$\qquad\qquad = (0_1 \bar{D}_4 0_5 \bar{D}_6 \bar{D}_8 0_9)$
$\qquad Df_1 = (G_2, G_3)$

Gate G_2. $\quad tc_2 = tc_1 \cap (\bar{D}_6 0_7 \bar{D}_{10})$
$\qquad\qquad = (0_1 \bar{D}_4 0_5 \bar{D}_6 0_7 \bar{D}_8 0_9 \bar{D}_{10})$
$\qquad Df_2 = (G_3, G_4)$

Gate G_3. $tc_3 = tc_2 \cap (0_3\bar{D}_8\bar{D}_{11})$
$ = (0_1 0_3 \bar{D}_4 0_5 \bar{D}_6 0_7 \bar{D}_8 0_9 \bar{D}_{10} \bar{D}_{11})$
$ Df_3 = (G_5 (2))$

Gate G_4. $tc_4 = tc_3 \cap (\bar{D}_{10}\bar{D}_{11}D_{12})$
$ = (0_1 0_3 \bar{D}_4 0_5 \bar{D}_6 0_7 \bar{D}_8 0_9 \bar{D}_{10} \bar{D}_{11}D_{12})$
$ Df_4 = (G_5)$

Gate G_5. $tc_5 = tc_4 \cap (0_9 D_{12}D_{13})$
$ = (0_1 0_3 \bar{D}_4 0_5 \bar{D}_6 0_7 \bar{D}_8 0_9 \bar{D}_{10} \bar{D}_{11}D_{12}D_{13})$
$ Df_5 = \emptyset$

Comment: The D-drive operation has proved successful for this path choice, and the last stage of the D-algorithm is to check the consistency of the coordinates that have been assigned either 0 or 1. This is achieved in a systematic fashion, working back from the highest numbered coordinate to the primary inputs. The terminal numbering convention, mentioned previously (section A1.1.1), ensures that no gate G_i is considered before another gate G_j if the output of G_i subsequently becomes an input to G_j.

The basic process is to select the highest numbered coordinate and intersect the test cube with a cube in a singular cover that:

(i) has this coordinate as an output coordinate, and
(ii) has the same output value as that in the test cube.

The intersection rules are the same as the D-drive rules (Table A1.1).

In the consistency operation, therefore, we have a list of coordinate positions whose value is fixed and which must be checked. This list is called the *fixed-value* list and is denoted by fv with an appropriate subscript. The positions are listed in descending numerical order.

The starting point for the consistency operation for this example is therefore defined by:

$tc_5 = (0_1 0_3 \bar{D}_4 0_5 \bar{D}_6 0_7 \bar{D}_8 0_9 \bar{D}_{10} \bar{D}_{11}D_{12}D_{13})$
$fv_5 = (9, 7, 5, 3, 1)$

The various stages in consistency make use of the singular cover data in Table A1.3 and are detailed below.

Coordinate 9. $tc_6 = tc_5 \cap (0_4X_50_9)$
                $= \emptyset$, try again
                $tc_6 = tc_5 \cap (X_40_50_9)$
                $= (0_10_3\bar{D}_40_5\bar{D}_60_7\bar{D}_80_9\bar{D}_{10}\bar{D}_{11}D_{12}D_{13})$
                $fv_6 = (7, 5, 3, 1)$

Comment: The first intersection failed (because $0_4 \quad \bar{D}_4 = \phi$). Fortunately another singular cover cube existed that satisfied the 0_9 condition.

Coordinate 7. $tc_7 = tc_6 \cap (0_20_7)$
                $= (0_10_20_3\bar{D}_40_5\bar{D}_60_7\bar{D}_80_9\bar{D}_{10}\bar{D}_{11}D_{12}D_{13})$
                $fv_7 = (5, 3, 2, 1)$

Comment: This intersection was successful but specified another fixed value on coordinate 2. This coordinate position has been added to the fixed value list.

Coordinate 5. $tc_8 = tc_7 \cap (0_20_5)$
                $= (0_10_20_3\bar{D}_40_5\bar{D}_60_7\bar{D}_80_9\bar{D}_{10}\bar{D}_{11}D_{12}D_{13})$
                $fv_8 = (3, 2, 1)$

Comment: There was no change in the test cube itself but the intersection confirmed that the fixed value on coordinate 5 was constant. At this stage, the fixed value vector only contains positions directly controllable by the primary inputs and the consistency operation is complete. The final test cube tc_8 can now be interpreted as a test for the initial fault and, in fact, defines $x_1 = x_2 = x_3 = 0$ as the test input, with $z_1 = 1$ if the fault is not present and $z_1 = 0$ if c_1 is s-a-1.

Application 2. Test for c_{11} s-a-0

To consolidate our understanding of the D-algorithm processes, Table A1.4 summarizes the D-drive and consistency operations for the source fault c_{11} s-a-0. The main point about this example is that it demonstrates how the algorithm handles possible choices situations. There are in general five types of choices:

Source Fault: c_{11} s-a-0

Choice points

Failure D-cubes: $(1_3 0_8 D_{11})$, $(0_3 1_8 D_{11})$ or $(1_3 1_8 D_{11})$

Select $(1_3 1_8 D_{11})$ arbitrarily   A

Path Choice: Only one possibility - G4,G5

D-DRIVE	Test Cube	D-Frontier	
Initial Condition	$tc_0 = (1_3 1_8 D_{11})$	$Df_0 = (G4)$	
Gate G4	$tc_1 = (1_3 1_8 1_{10} D_{11} \bar{D}_{12})$	$Df_1 = (G5)$	
Gate G5	$tc_2 = (1_3 1_8 0_9 1_{10} D_{11} \bar{D}_{12} \bar{D}_{13})$	$Df_2 = \emptyset$	

CONSISTENCY	Test Cube	Fixed-Value	
Initial condition	$tc_2 = (1_3 1_8 0_9 1_{10} D_{11} \bar{D}_{12} \bar{D}_{13})$	$fv_1 = (10,9,8,3)$	
Co-ordinate 10	$tc_3 = (1_3 1_6 X_7 1_8 0_9 1_{10} D_{11} \bar{D}_{12} \bar{D}_{13})$	$fv_2 = (9,8,6,3)$	B
Co-ordinate 9	$tc_4 = (1_3 0_4 X_5 1_6 X_7 1_8 0_9 1_{10} D_{11} \bar{D}_{12} \bar{D}_{13})$	$fv_3 = (8,6,4,3)$	C
Co-ordinate 8	$tc_5 = (1_1 1_3 0_4 X_5 1_6 X_7 1_8 0_9 1_{10} D_{11} \bar{D}_{12} \bar{D}_{13})$	$fv_4 = (6,4,3,1)$	
Co-ordinate 6	$tc_6 = (1_1 1_3 0_4 X_5 1_6 X_7 1_8 0_9 1_{10} D_{11} \bar{D}_{12} \bar{D}_{13})$	$fv_5 = (4,3,1)$	
Co-ordinate 4	$tc_7 = \emptyset$ $(1_1 \quad 0_1 = \emptyset)$		
Consistency Failure - Back up to Point C and try again			
Co-ordinate 9	$tc_4 = (1_3 X_4 0_5 1_6 X_7 1_8 0_9 1_{10} D_{11} \bar{D}_{12} \bar{D}_{13})$	$fv_3 = (8,6,5,3)$	
Co-ordinate 8	$tc_5 = (1_1 1_3 X_4 0_5 1_6 X_7 1_8 0_9 1_{10} D_{11} \bar{D}_{12} \bar{D}_{13})$	$fv_4 = (6,5,3,1)$	
Co-ordinate 6	$tc_6 = (1_1 1_3 X_4 0_5 1_6 X_7 1_8 0_9 1_{10} D_{11} \bar{D}_{12} \bar{D}_{13})$	$fv_5 = (5,3,1)$	
Co-ordinate 5	$tc_7 = (1_1 0_2 1_3 X_4 0_5 1_6 X_7 1_8 0_9 1_{10} D_{11} \bar{D}_{12} \bar{D}_{13})$	$fv_6 = (3,2,1)$	

End of consistency, fv_6 = (primary inputs only)

Test given by $x_1 = 1$, $x_2 = 0$, $x_3 = 1$, $z_1 = 0$

TABLE A1.4 Application of D-Algorithm for $c_{11}/0$

(i) choice of a failure D-cube covering the source fault;

(ii) choice of a path or set of paths;

(iii) choice of propagating D-cubes in the D-drive process;

(iv) choice of non-propagating D-cubes for D-blocking purposes; and

(v) choice of singular cover cubes during the consistency process.

In Table A1.4 there are three choice points, labelled A, B, and C; point A is concerned with the choice of failure D-cube and points B and C occur during the consistency process. Further down the consistency stages, we see that the test cube reduces to the empty cube ϕ and, at this point, the algorithm "backs up" to the last choice point, selects one of the other choices, and tries again. In this example, the second choice at point C leads to a successful result. If this had not been the case, the algorithm would have backed up to point B, selected another singular cover cube, and tried again. If this had failed, the algorithm would have gone right back to point A and tried again with a different failure D-cube.

In general, therefore, a record of all choices is kept and, in the limit, if the algorithm continually terminates unsuccessfully, all possibilities are tried. It is precisely this feature of the D-algorithm that makes it an algorithm.* A full implementation on a digital computer therefore must contain the mechanisms for remembering the position of choice points and which choices were made, and for restoring the test cube, with its associated D-frontier or fixed value list, back to the state just before the choice point if necessary.

A1.1.6 The LASAR algorithm

Before we consider the formal procedures for generating D-cube data, we should comment on an important variation of the D-algorithm. This is the "critical path" technique, more commonly referred to as the LASAR algorithm. (The Logic Automated Stimulus And Response system is a well-known commercial system for generating and evaluating test patterns.) The LASAR algorithm can be described in terms of the D-cube concepts and notation but the strategy for test generation is different to that of the D-algorithm. Essentially, the algorithm attempts to create sensitive paths by working back from an assigned value on a primary output. In following a route back through the circuit, an attempt is made to assign logic values to the input side of a device such that at least one input can control the output, i.e., is a fault generator. Where a choice of assignments exists, the selection can either be arbitrary or it can be one that minimizes the number of sensitive inputs. In practice, the LASAR algorithm seeks to identify a single sensitive path (one control input) rather than a multiple set of sensitive paths. In this way the possibility of problems caused by reconvergence are reduced. As with the D-algorithm, however, choice situations do arise and it is necessary to remember what the choices were and which one was selected in order to back-track if necessary.

*The word *algorithm* generally means a procedure that is described by a decision tree that is directed, bounded, and terminated and which gives correct answers for the specified range and type of data input.

The procedure will be illustrated on the example circuit shown in Fig. A1.5.

Step 1 Assign $c_{13} = 1$

Step 2 Select an input assignment to G_5 that satisfies $c_{13} = 1$ and that sets just one input to a control (sensitive) status.
Comment: Two possibilities exist—either $c_9 = 1$ and $c_{12} = 0$ or $c_9 = 0$ and $c_{12} = 1$. The $c_9 = 1$, $c_{12} = 1$ assignment does not produce a sensitive input. Arbitrarily, we will select $c_9 = 1$ and $c_{12} = 0$. The sensitive input is c_9.

Step 3 Follow the critical path back and assign values to the inputs of G_1.
Comment: There is only one possibility here: $c_4 = 1$ and $c_5 = 1$. Both these values are necessary, both are sensitive, and both originate from primary inputs. The result of this assignment therefore is to make $x_1 = 1$, $x_2 = 1$.

Step 4 Check the consistency of the non-sensitive assignment to $c_{12}(=0)$.
Comment: The only possibility here is to assign $c_{10} = 1$ and $c_{11} = 1$.

Step 5 Check the consistency of the non-sensitive assignment to $c_{10}(=1)$.
Comment: This assignment is already satisfied by the earlier assignments to x_1 and x_2. Note, however, that we must check that the effect of either x_1 or x_2 changing does not invalidate the $c_{10} = 1$ assignment. Both x_1 and x_2 are sensitive and, if x_1 changes, say, then all nodes connected to x_1 will change. This includes c_6. Similarly, a change to x_2 will affect c_7. If we assume single fault only, then it is not possible for both c_6 and c_7 to change simultaneously. The integrity of $c_{10} = 1$ is preserved, therefore. If it were discovered that a single fault can cause both c_6 and c_7 to change, then this would be classed as an inconsistency and an alternative possibility tried.

Step 6 Check the consistency of the non-sensitive assignment to $c_{11}(=1)$.
Comment: At first sight, it would appear that this requirement is

already met by the $x_1 = 1$ assignment, and that the other input to G_3 (c_3) can be left unassigned. From the comments made at the previous step, however, we see that c_3 must be assigned 1 in order to preserve $c_{11} = 1$ should the value of c_8 change. This consideration produces $x_3 = 1$.

The process of generating a test for the initial assignment of $c_{13} = 1$ is now complete. The test is $x_1 = 1, x_2 = 1, x_3 = 1$ and the critical paths are (c_1-c_4-c_9-c_{13}) and (c_2-c_5-c_9-c_{13}). The whole procedure is now repeated starting with $c_{13} = 0$.

In practice, the processes of backward trace of critical paths and non-sensitive paths are more integrated than was suggested by the example. Also, the method suffers from the fact that it seeks to generate single paths only. This may mean that tests for certain faults are never found if their detection relies on multiple paths. Another objection is that only two tests per output are generated: one with the output at 1 and the other with the output at 0. For the example above, we already know that at least four tests are required to provide full fault-cover of all fanout trunk and single nodes s-a-1 and s-a-0. In this respect, therefore, the fault-cover of the two tests generated by the LASAR algorithm must be incomplete. In practice, the algorithm can be repeated for the same initial value on the primary output but with a directive to explore alternative input assignments, where they exist.

This concludes the description of the D-algorithm and its derivative, the LASAR algorithm. The next section describes the formal procedures for generating D-cube data.

A1.2 Derivation of D-cubes

Now that the process of intersection has been defined, we are able to show how propagating, non-propagating, and failure D-cubes are derived formally. These cubes are essential data for the application of the D-algorithm to test pattern generation problems. The terms have already been defined and illustrated in the previous sections, in which the cubes for a 2-input NAND gate were derived in an intuitive fashion. Table A1.2 presented a list of all such cubes for all basic logic gates. In this section, the formal procedures for deriving the cubes are presented and illustrated by application to the same gate. Generalizations to n-input gates where n > 2 are easily made. Fig. A1.6 shows the gate and fault-free truth table.

A1.2.1 Propagating D-cubes

The essence of a propagating D-cube for a logic element is that it identifies an input or set of inputs whose value, 0 or 1, controls the value of the output. The simplest way to derive these, therefore, is by studying the truth table of the element and, in particular, by comparing pairs of inputs whose outputs differ.

In Fig. A1.6(b), each cube (row) of the truth table is either a member of the ON-array A_1 if the output value is 1, or a member of the OFF-array A_0 if the output value is 0. Also, each cube has been identified by a, b, c or d.

Therefore:

$$A_1 = \{a, b, c\}$$
$$A_0 = \{d\}$$

Now consider cubes c and d. As far as the input coordinates are concerned, c_1 has the same value (1) whereas the values on c_2 differ. The value on the output coordinate also differs. What this says, therefore, is that if c_1 is held at 1, the value of the output coordinate c_3 is directly controlled by the value on the other input coordinate c_2. This then defines a propagation D-cube and, if we adopted the arbitrary conven-

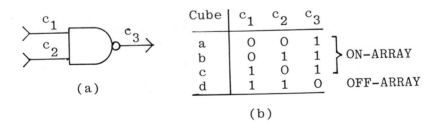

Cube	c_1	c_2	c_3	
a	0	0	1	ON-ARRAY
b	0	1	1	
c	1	0	1	
d	1	1	0	OFF-ARRAY

(a)   (b)

FIGURE A1.6 2-Input Nand Gate

tion that a $0 \to 1$ change from the first cube to the second be written D, with $\bar{D}$ representing the $1 \to 0$ change, we arrive at the propagation D-cube $(1_1 D_2 \bar{D}_3)$. Because the D, $\bar{D}$ convention was arbitrary, we could have chosen the opposite and derived the propagation D-cube as $(1_1 \bar{D}_2 D_3)$, i.e., the dual.

What, in effect, has happened is that we have intersected the two cubes c and d to form a propagation D-cube result. This procedure can be formalized in the following way.

Let α and β be two cubes of length n selected from the element truth-table, such that

$$\alpha = (\alpha_1, \alpha_2, \ldots, \alpha_i, \ldots, \alpha_n)$$

$$\beta = (\beta_1, \beta_2, \ldots, \beta_i, \ldots, \beta_n)$$

and $\alpha_i, \beta_i = 0, 1$ for all i.

The intersection $\alpha \cap \beta$ forms a propagation D-cube provided:

(i) at least one of the output coordinates in α differs in value from the corresponding coordinate in β. (This is the general condition. For single output elements, this requires α to come from the ON-array A_1 and β from the OFF-array A_0, or vice versa.)

(ii) the following intersection rules are used:

$$\alpha_i \cap \beta_i = \alpha_i \text{ if } \alpha_i = \beta_i$$
$$= D \text{ if } \alpha_i = 0 \text{ and } \beta_i = 1$$
$$= \bar{D} \text{ if } \alpha_i = 1 \text{ and } \beta_i = 0$$

Exhaustive application of this procedure to the ON- and OFF-array cubes for the 2-input NAND gate (Fig. A1.6(b)) produces the following propagation D-cubes:

$$a \cap d = D_1 D_2 \bar{D}_3$$

$$b \cap d = D_1 1_2 \bar{D}_3$$

$$c \cap d = 1_1 D_2 \bar{D}_3$$

The duals are derived in the usual way.

A1.2.2 Non-propagating D-cubes

The derivation of non-propagating D-cubes is very similar to that of propagating D-cubes. The only difference is that the non-propagating D-cube defines the condition under which a change on one or more of the inputs does *not* produce a corresponding change on the output (or outputs if the element is multi-output). This means that non-propagating D-cubes are formed by intersecting cubes with the same output values.

Put more formally:

Let α and β to two cubes of length n selected from the element truth-table such that:

$$\alpha = (\alpha_1, \alpha_2, \ldots, \alpha_i, \ldots, \alpha_n)$$
$$\beta = (\beta_1, \beta_2, \ldots, \beta_i, \ldots, \beta_n)$$
$$\text{and } \alpha_i, \beta_i = 0, 1 \text{ for all } i$$

The intersection $\alpha \cap \beta$ forms a non-propagating D-cube provided:

(i) the value of the output coordinates in α are identical to those in β, and

(ii) the following intersection rules are used:

$$\alpha_i \cap \beta_i = \alpha_i \text{ if } \alpha_i = \beta_i$$
$$= D \text{ if } \alpha_i = 0 \text{ and } \beta_i = 1$$
$$= \bar{D} \text{ if } \alpha_i = 1 \text{ and } \beta_i = 0$$

Exhaustive application of this procedure to the 2-input NAND gates cubes produces the following non-propagating D-cubes.

$$a \cap b = 0_1 D_2 1_3$$
$$a \cap c = D_1 0_2 1_3$$
$$b \cap c = D_1 \bar{D}_2 1_3$$

As before, the duals follow directly.

A1.2.3 Failure D-cubes

Failure D-cubes are also formed by cube intersection, but this time, one of the cubes is selected from the fault-free truth table and the other from the truth table relating to the faulty behavior. To illustrate the procedure, we will derive the failure D-cube for a s-a-1 fault on connection c_2 of the 2-input NAND gate. Table A1.5 shows the truth tables corresponding to the fault-free and faulty behavior of the gate and it is easily seen that the only test condition is defined by $c_1 = 1$, $c_2 = 0$. Furthermore, the fault-free output is 1, becoming 0 if the fault is present. This, then, defines the failure D-cube $1_1 0_2 D_3$ for the fault c_2 s-a-1 (recall that the convention for interpreting D's in failure D-cubes is $D(\bar{D}) = 1$ (0) if the fault is not present and $D(\bar{D}) = 0$ (1) if the fault is present).

The general procedure for finding failure D-cubes, therefore, is to search down the output values corresponding to the fault-free and faulty behavior truth tables. If a difference of output value is found, the failure D-cube is defined by the input coordinate values together with the D or $\bar{D}$ assignment on the output coordinate. For multi-output elements it is necessary that at least one output coordinate differs in value. Those outputs that do not differ are left at their fixed 0 or 1 value in the failure D-cube.

This procedure for deriving failure D-cubes is quite general. The following two examples demonstrate the derivation of failure D-cubes for failures other than those of s-a-1 or s-a-0 type. The first considers a short circuit between two connections, the second a wired-OR short circuit fault that causes a transformation of the function performed by a 3-input NAND gate.

c_1	c_2	c_3(fault-free)	c_3(c_2s-a-1)
0	0	1	1
0	1	1	1
1	0	1	0
1	1	0	0

TABLE A1.5 NAND Gate Truth-Tables

A1.2.4 Short-circuit between two connections

Fig. A1.7(a) shows two connections, c_1 to c_1^* and c_2 to c_2^*, and Fig. A1.7(b) shows the same two connections with a short circuit, assumed to be of the wired-OR type. We can consider the two connections to be a 2-input, 2-output "element" and derive truth tables corresponding to the fault-free and faulty behavior. These truth tables are also shown in Fig. A1.7.

Comparing individual rows of the two truth tables produces the following two failure D-cubes:

c_1	c_2	c_1^*	c_2^*
0	1	D	1
1	0	1	$\bar{D}$

The reader should verify that if the short circuit had been assumed to be of the wired-AND type, the corresponding failure D-cubes would be given by:

c_1	c_2	c_1^*	c_2^*
0	1	0	D
0	1	D	0

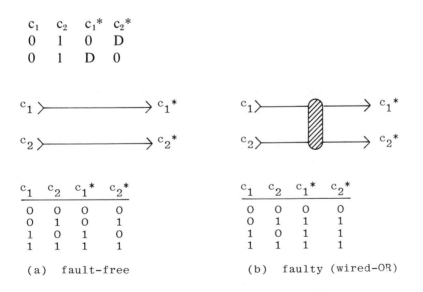

c_1	c_2	c_1^*	c_2^*
0	0	0	0
0	1	0	1
1	0	1	0
1	1	1	1

(a)   fault-free

c_1	c_2	c_1^*	c_2^*
0	0	0	0
0	1	1	1
1	0	1	1
1	1	1	1

(b)   faulty (wired-OR)

FIGURE A1.7 Short-Circuit Failure Between Two Connections

Logic function transformation

Fig. A1.8(a) shows a 3-input NAND gate with a wired-OR short circuit fault between two of the input connections. (This type of fault is called an *input bridging fault*.) The fault-free and faulty behavior truth tables are shown in Fig. A1.8(b). The failure D-cubes are found to be:

$$0_1 \quad 1_2 \quad 1_3 \quad D_4$$
$$1_1 \quad 0_2 \quad 1_3 \quad D_4$$

(a) wired-OR input
     bridging fault.

c_1	c_2	c_3	c_4 (fault-free)	c_4 (faulty)
0	0	0	1	1
0	0	0	1	1
0	1	0	1	1
0	1	1	1	0
1	0	0	1	1
1	0	1	1	0
1	1	0	1	1
1	1	1	0	0

(b) Truth-tables

FIGURE A1.8 Input Bridging Fault on NAND Gate

A1.3 Historical Development of the D-algorithm

The concepts of a sensitive path can be traced to 1959 and Eldred's work [1] on the Datamatic-1000 processor. Eldred introduced the notion of certain inputs controlling the output of a gate and used this concept to propagate fault effects from OR gates through successor AND gates and thence to observable flip-flops. More details can be found either in Eldred's paper or in the books by Chang, Manning, and Metze [2, Chapter 3], or Breuer [3, Chapter 7].

Armstrong's paper in 1966 [4] was notable for its formalization of the sensitive path technique. In particular, he developed an algebraic expression for the function of a circuit working from the structure. This expression was termed the *equivalent normal form* (enf) and its main feature was a retention of structural information. A list of gates contributing toward each term was attached to the term itself and Armstrong then showed how the value of the output function could be made dependent on just one of the terms. This was achieved by assigning values of 1 to the literals in the term chosen and values of 0 to at least one of the literals in each of the other terms in the expression. (This is not always possible, of course.) In effect, therefore, since each term identifies a path through the circuit, the procedure determines all the possible single sensitive paths.

The fact that the procedure does not identify parallel path sets is its main shortcoming. Armstrong was aware of parallel path possibilities but he did not generalize the enf technique to encompass parallel paths. He did, however, discuss fault-effect reconvergence and identified the two possibilities of positive and negative reconvergence (he called these Case 1 and Case 2, respectively). Descriptions of the enf technique can be found in Friedman and Menon [5], Kohavi [6], or Lee [19].

Because of the single path nature of the enf technique, it is now called a *1-dimensional* path sensitizing technique. The first truly *n-dimensional* technique was the D-algorithm, and this was initially described in a paper by Roth in 1966 [7]. This paper made use of and built on a number of concepts developed by Roth in earlier papers. These included the use of a cubical notation and the process of cube intersection. The 1966 paper describes the D-algorithm, later referred to as DALG-I, and proves the general theorem that if a test for a particular fault exists, then the D-algorithm will find it.

The algorithm was subsequently updated to a new version, DALG II, described in the paper by Roth et al. [8]. The main difference between DALG-I and DALG-II is that whereas DALG-I completed the D-drive before starting the consistency operation, DALG-II intermixes these operations, i.e., as soon as a fixed 0 or 1 value is assigned to some coordinate in the test cube during the D-drive process, the consistency of the assignment is checked immediately, going back as far as possible. It was claimed that this is more efficient, in that inconsistencies are discovered earlier, thereby saving unnecessary computation. (This claim has subsequently been upheld by other research workers.)

The other interesting feature of the DALG-II paper is that it applies the D-algorithm to a circuit containing a fault that can only be tested by establishing a parallel path from the site of the fault to the primary output. The circuit was originally described by Schneider [9] and is often used as an example circuit for illustrating the processes of the D-algorithm (see [2, 5], for instance).

Other papers on the D-algorithm have appeared quite regularly since the original disclosure in 1966. Some are by Roth and his co-workers, whereas others come from other sources [21–25]. Most of these papers deal explicitly with applications of the D-algorithm to sequential circuits; comment on this subject is made shortly. Details of the LASAR algorithm can be found in [18, Chap. 2], or in the original papers by Thomas [26], or Bowden [27].

Finally, comment should be made on the differences between the description of the D-algorithm in this appendix and descriptions that exist elsewhere, notably in Roth's paper [7] and in other text books [2,3,5]. The comments are as follows:

(i) In this appendix, we have chosen to base the procedure for deriving propagation and failure D-cubes on the full version of the truth table (section A1.2). The alternative is to use the more compact form of the truth table, i.e., the singular cover. An algorithm based on this form is described in [7] and is slightly more complicated because of the possibility of an X in the table.

(ii) The original rules for D-cube intersection presented in [7] are also more complicated than those presented in section A1.1.4. The reason for this is that no use was made of the dual version of the propagating D-cubes. This gave rise to specific intersection possibilities that were indeterminate, e.g., D $\bar{D}$ could lead to a result of D if it was subsequently found that the $\bar{D}$ term could be inverted back to D without violation of the logic conditions. An explicit use of the dual propagating D-cubes avoids this problem, as is indicated in a later paper on the D-algorithm [10].

(iii) The concept and use of non-propagating D-cubes has not been pursued as strongly as it might in other descriptions of the D-algorithm. As we have seen, however, the ability to block the propagation of fault-effects is just as important as the ability to propagate the fault-effect and, in this appendix, non-propagating D-cubes have been treated as equally important as propagating D-cubes.

(iv) The original papers on the D-algorithm are relatively mathematical in their description, and are based on the "calculus of D-cubes.'' This calculus is introduced and defined in earlier publications of Roth [1,2]; a useful summary of much of the terminology can be found in [13, Chapter 7].

A1.4 Comment on Extending the D-algorithm to Handle Circuits Containing Stored-State Devices and Global Feedback

There are two additional and important factors to be considered when applying the concepts of path sensitization to logic circuits containing stored-state devices such as flip-flops and possibly global feedback (such circuits will be referred to, rather loosely, as sequential circuits). The first is that the progressive propagation of fault-effects along a sensitive path that includes stored-state devices requires an additional timing parameter associated with a test cube to identify the relative timing of one cube with another. It is also necessary to initialize a stored-state device into a known start state before it cna be used as a fault transmitter for a sensitive path coming in on a particular input.

The second factor is the possible presence of global feedback, i.e., feedback from one part of a circuit to another part wherein the value fed back can ultimately affect the original value itself. If this is structurally possible, it may well mean that what was required to be a fixed-value, and specified as such at a particular timeslot, is subsequently ''overwritten'' at a later timeslot as the postulated fault-effect (i.e., the D's in the circuit) are driven around the loop.

These two factors taken together have served as major barriers to the successful extension of algorithms, like the D-algorithm, to sequential circuits. Various attempts have been made (see Kubo [14], Putzolu and Roth [10], Bennetts [15], Muth [16], for example) but the conclusion has been, in general, that it is very easy to design a circuit that defeats the algorithm! Unfortunately, logic designers are very skilled at this and the state-of-the-art is that there is still no general-purpose algorithm that suits complex sequential circuits (complex in terms of the degree and number of structural feedback paths and in terms of the use of stored-state devices, both stable and unstable, such as monostables). Indeed, current thinking is to constrain the design of sequential circuits in such a way as to enable the application of limited versions of test-pattern generation algorithms, and this has given rise to the principle of designing a circuit to be testable in the first place. The most well-known examples of this, of course, are IBM's Level-Sensitive Scan Design (Chap. 9) and NEC's Scan-Path [21,24,25]. Both these techniques are particular implementations of the Scan-In, Scan-Out principle.

In summary, therefore, the consensus of opinion is to constrain the circuit to fit the abilities and limitations of the test-pattern generation algorithm rather than to attempt to cover all possible eventualities. In theory, this is not difficult. In practice, only enlightened logic designers acknowledge that the circuits they design must eventually be tested.

A1.5 References and Bibliography

1. R. C. Eldred. *Test Routines Based on Symbolic Logic Statements*. Journal ACM, Vol.6, No.1, January 1959, pp.33–36.

2. H. Y. Chang, E. G. Manning, and G. Metze. *Fault Diagnosis of Digital Systems*. Wiley Interscience, 1970.

3. M. A. Breuer (Ed.). *Design Automation of Digital Systems, Volume 1, Theory and Techniques*. Prentice-Hall, 1972.

4. D. B. Armstrong. *On Finding a Nearly Minimal Set of Fault Detection Tests for Combinational Logic Nets*. IEEE Trans. Electronic Computers, Vol.EC-15, No.1, January 1966, pp. 63–73.

5. A. D. Friedman and P. R. Menon. *Fault Detection in Digital Circuits*. Prentice-Hall, 1971.

6. Z. Kohavi. *Switching and Finite Automata Theory*. McGraw-Hill Computer Science Series, 1970.

7. J.P. Roth. *Diagnosis of Automata Failures: A Calculus and a Method*. IBM Journal, Vol.10, No.7, July 1966, pp. 278–291.

8. J. P. Roth, W. G. Bouricius, and P. R. Schneider. *Programmed Algorithms to Compute Tests to Detect and Distinguish Between Failures in Logic Circuits*. IEEE Trans. Electronic Computers, Vol.EC-16, No.5, October 1967, pp. 567–580.

9. P. R. Schneider. *On the Necessity to Examine D-chains in Diagnostic Test Generation–an Example*. IBM Journal, Vol.11, No.1, January 1967, p. 114.

10. G. R. Putzolu and J. P. Roth. *A Heuristic Algorithm for the Testing of Asynchronous Circuits*. IEEE Trans. Computers, Vol.C-20, No.6, June 1971, pp. 631–647.

11. J. P. Roth. *Minimisation over Boolean Trees*. IBM Journal, Vol.4, No.15, November 1960, pp. 543–558.

12. J. P. Roth and R. M. Karp. *Minimisation over Boolean Graphs*. IBM Journal, Vol.6, No.2, April 1962, pp. 227–238.

13. F. J. Hill and G. R. Peterson. *Introduction to Switching Theory and Logical Design*. Wiley, 1968.

14. H. Kubo. *A Procedure for Generating Test Sequences to Detect Sequential*

Circuit Failures. NEC Journal of R and D, No.12, 1968, pp. 66–78.

15. R. G. Bennetts et al. *A Modular Approach to Test Sequence Generation for Large Digital Networks.* Digital Processes, Vol.1, 1975, pp. 3–24.

16. P. Muth. *A Nine-valued Circuit Model for Test Generation.* IEEE Trans. Computers, Vol.C-25, 1976, pp. 630–636.

17. D. W. Lewin. *Computer Aided Design of Digital Systems.* Arnold, 1977.

18. M. A. Breuer and A. D. Friedman. *Diagnosis and Reliable Design of Digital Systems.* Computer Science Press, 1976.

19. S. C. Lee. *Digital Circuits and Logic Design.* Prentice-Hall, 1976.

20. S. C. Lee. *Modern Switching Theory and Digital Design.* Prentice-Hall, 1978.

21. K. Tomita et al. *Test Generation System for Digital Circuits.* NEC Research and Development, No.49, April 1978, pp. 16–24.

22. M. A. Breuer and A. D. Friedman. *Test/80: A Proposal for an Advanced Automatic Test Generation System.* Proc. IEEE Autotest Conf., 1979, pp. 305–312.

23. M. A. Breuer and A. D. Friedman. *Functional Level Primitives in Test Generation.* IEEE Trans. Computers, Vol.C-29, No.3, March 1980, pp. 223–234.

24. A. Yamada et al. *Automatic Test Generation for Large Digital Circuits.* Proc. IEEE 14th Design Automation Conf., 1977, pp. 78–83.

25. K. Kani et al. *CAD in the Japanese Electronics Industry.* Proc. EEC Symposium on CAD of Digital Electronic Circuits and Systems, 1978, pp. 115–131.

26. J. J. Thomas. *Automated Diagnostic Test Programs for Digital Networks.* Computer Design, August 1971, pp. 63–67.

27. K. R. Bowden. *A Technique for Automatic Test Generation for Digital Circuits.* Proc. IEEE Intercon, 1975, Session 15, pp. 1–5.

Index

Note that this index does not contain a reference to those items listed in Chapter 10.

Abbreviations

ATE	Automatic Test Equipment
ATPG	Algorithmic Test Pattern Generation
ATS	Automatic Test System
B-U-T	Board-Under-Test
CMOS	Complementary Metal Oxide Semiconductor
DA	IEEE Design Automation conference
DAFTC	Journal of Design Automation and Fault-Tolerant Computing
Df	D-frontier
DMA	Direct Memory Access
D/S	Driver/Sensor
D-U-T	Device-Under-Test
EAROM	Electrically-Alterable Read-Only Memory
ECL	Emitter-Coupled Logic
enf	equivalent normal form
FC	Fault Cover
FTCS	IEEE Fault-Tolerant Computing Symposium
fv	fixed value
IC	Integrated Circuit
IEEETC	IEEE Transactions on Computers
IEEETEC	IEEE Transactions on Electronic Computers
I/O	Input/Output
KGB	Known Good Board
LSI	Large-Scale Integration
LSSD	Level-Sensitive Scan Design
MC	Microcomputer